Designermonographien 5
Erweiterte Neuauflage

Designer Monographs 5
Expanded New Edition

Peter Maly

Peter Maly

Designermonographien / Designer Monographs 5
Hrsg. von / Edited by Alex Buck & Matthias Vogt

Josef Kremerskothen
Helmut Lübke
Wolfgang Nagel
Verner Panton
Peter Zec
Petra Schwab

verlag form

Inhalt Contents

Vorwort

Über Peter Maly:

Stellen Sie sich vor, Sie befinden sich in einer Wohnung und alles erscheint Ihnen wunderbar stimmig: zeitgemäß, ohne bloß modisch zu wirken; klassisch, jedoch nicht antiquiert; präsent, aber dennoch dezent; qualitativ hochwertig, ohne aufdringlich zu sein.

Die Sitzmöbel, Tische und Stühle, die Schränke und Kommoden, die Teppiche und Leuchten, alle Gegenstände – obwohl augenscheinlich nicht zum selben Zeitpunkt erworben und auch ohne vordergründig gestalterische Ambitionen kombiniert – harmonieren zwanglos miteinander.

Was aber verbindet die unterschiedlichen Objekte? Offensichtlich ist es die Qualität, Bedürfnisse zu reflektieren, die dem Benutzer zuvor nicht bewußt waren. Ja, die Gegenstände dieser Wohnung ergänzen einander in ihrer Wirkung und ergeben einen sympathischen Kanon von Bequemlichkeit und zurückhaltender Eleganz.

Fragen nach der Herkunft der Gegenstände oder nach ihrem Gestalter stellen sich zunächst nicht, handelt es sich hier doch um Objekte, von denen man annimmt, sie wären immer schon dagewesen, gehörten zu einem kulturellen Fundus, wären Ergebnis einer Art ästhetischen Generationenvertrags.

Nicht zuletzt diese unaufdringliche Qualität, den Gastgeber nicht gleich ansprechen zu müssen und nach dem »Woher« und »Von wem« der Gegenstände zu befragen, ist angenehm unzeitgemäß und deshalb so sympathisch. Stellte man diese Fragen dennoch, erhielte man vielleicht gar keine Antwort. Ohnehin ist die Chance, daß der Be-

Preface

About Peter Maly:

Imagine you are standing in a room and everything strikes you as being perfectly in order; contemporary without seeming trendy, classic but not antiquated. The room has a strong presence that is nonetheless discreet, and is of a high quality without being gaudy.

The seating, tables and chairs, cupboards and chests of drawers, the rugs and lamps – indeed all the objects in the room – are in harmony, although they were evidently not all designed at the same time, and were combined without any obvious underlying artistic ambitions.

But just what links these disparate objects? It is clearly the ability to reflect needs which were previously unknown to the users themselves. Indeed, the objects in this room complement one another and create a sympathetic vocabulary of comfort and elegant reserve.

Questions as to the origin of the objects or who they were designed by do not initially spring to mind, for you assume that these objects have always been here, that they belong to the basic stocks of interior design, as it were, or are the result of some kind of aesthetic agreement across the generations.

Today, it is not least this unobtrusive quality, the fact that you do not have to address your host directly and ask »where« and »by whom« these objects were created, which is so pleasantly untypical, and for that very reason so appealing. In any case, if you were to ask, you might not even receive an answer, as the person you were asking would be unlikely to know the name of the designer. But precisely when juxtaposed to everyday culture today, it is this unobtrusive feel to the works of designer and interior designer Peter Maly that is so fascinating, that makes them seem the obvious choice.

fragte den Namen des Designers kennt, relativ gering. Aber gerade das Unauffällig-Selbstverständliche vieler seiner Möbel macht bei der Auseinandersetzung mit der Alltagskultur unser Interesse am Designer und Innenarchitekten Peter Maly aus.
Hat schließlich der Betrachter die dezente Präsenz der Malyschen Objekte einmal erkannt, beginnt er sich für sie zu interessieren und nun doch all jene Fragen nach der Herkunft und dem Designer zu stellen. Aber diese zu beantworten, ist nicht Aufgabe des Vorworts.

Über das vorliegende Buch:

… ist das fünfte in einer Reihe von Monographien mit eigener Charakteristik.
Im Mittelpunkt stehen zentrale Ansätze oder Begriffe der Entwurfsphilosophie Peter Malys, welche von Autoren kommentiert werden, die in einem nahen, gleichwohl nicht vorbehaltlos unkritischen Verhältnis zum Protagonisten stehen.
Peter Maly selbst ist bei der Konzeption dieses Buches aktiv geworden: die mittleren sechzehn Seiten standen ihm zur freien Verfügung. Der Designer hat sie mit Collagen gefüllt, die auf die verschiedenen Jahrzehnte seines Schaffens Bezug nehmen. Auf diese Weise erhält das Buch eine Intimität, die die Person Peter Maly fast greifbar werden läßt.
Eine große Anzahl von Freunden, Zeitgenossen, Kritikern, Weggefährten, Kollegen und Produzenten wurde von uns nach seiner Person befragt; einige der Kommentare fanden Eingang in dieses Buch und ermöglichen so eine noch differenziertere Betrachtung.
Wenn sich auch viele der alten Weisheiten im Design überlebt haben bzw. als nicht mehr zeitgemäß ausgemustert wurden – die »Einheit von Inhalt und Form« hat als Qualität überlebt. So liegt auch diesem Buch – wie allen dieser Reihe – ein grafisches Konzept zugrunde, das dem Inhalt adäquat ist.

Über das Dankesagen:

… bleibt noch der Dank für Mitarbeit, Unterstützung und Diskussion an Karin Görling, Ines Wagemann und Thea Weber sowie an all diejenigen, die mit ihren Statements über Peter Maly zum Facettenreichtum dieses Buches beigetragen haben.

Alex Buck und Matthias Vogt

Möbel gehören zu den ältesten Kulturgütern der Menschheit, auch Stühle. Sinnvolle Designinnovationen sind hier sehr selten. Weil dies so ist, entfernen sich manche Entwerfer aus der strengen Welt des Designs, in der die Maße des menschlichen Körpers und die Funktion des Möbels rigide Regeln aufstellen. Sie schaffen Wohnobjekte, was selbst dann zulässig ist, wenn die Funktionalität der Produkte angezweifelt werden kann. Allerdings: In diesen Fällen unterliegt die Beurteilung solcher Objekte den Gesetzen der Kunst, nicht denen des Designs.
Und doch gibt es ein- oder zweimal in einem Jahrzehnt seltene Glücksfälle, wenn ein Design wirklich innovativ ist und ein Möbel gleichzeitig als Designobjekt und als Kunstobjekt bestehen kann. Ein solcher Glücksfall ist das Sitzmöbel »Zyklus« von Peter Maly, von dem heute die Fachwelt überzeugt ist: Der »Zyklus« ist das herausragende Möbel, das in Deutschland in den fünfzig Jahren seit Kriegsende entstanden ist. Es gehört in eine Reihe mit dem Wiener Kaffeehausstuhl von Michael Thonet, dem »Hohen« von Mackintosh, dem B3 »Wassily« von Marcel Breuer und dem »Lounge-Chair« von Charles & Ray Eames. Mit anderen Worten: Der »Zyklus« wird überdauern.

Furniture is one of humanity's oldest cultural goods. This includes chairs as well. Sensible design innovations in this area are extremely few and far between. As a result, a number of designers quit the stringent world of design, in which the proportions of the human body and the function of the furniture impose rigid rules. Instead, these designers create objects for the home, something which is permissible even if the functionality of the products is open to question. However, any assessment of these objects is subject to the laws of art, not design. And yet, there are still those rare strokes of luck, perhaps once or twice in a decade. And they occur when a design is truly innovative, and the furniture can hold its own both as a design object and an *objet d'art.* The »Zyklus« seating furniture by Peter Maly is this kind of lucky break, and has convinced the experts. »Zyklus« is the outstanding furniture to emerge in Germany in the 50 years since the war. It belongs next to works such as the Viennese coffee house chair by Michael Thonet, the »High-backed chair« by Mackintosh, the B3 »Wassily« by Marcel Breuer and the »Lounge Chair« by Charles and Ray Eames. To put it another way, »Zyklus« is a work of enduring value …

Erich Naumann

Hauptgeschäftsführer des Verbands der Deutschen Möbelindustrie in Wiesbaden

Chief Director of the German Furniture Industry Association in Wiesbaden, Germany

Once the observer has recognised the unobtrusive aura of Maly's objects, he gradually becomes interested in them, and begins asking questions such as »where« and »by whom« after all. Providing the answers, however, is not within the scope of the introduction.

About this book:

As the fifth volume in the monograph series, this book has its own special character, for it focuses on the key methods and concepts of Peter Maly's design philosophy. A number of people who are well-acquainted with the protagonist offer commentary on his work, albeit not always merely uncritical praise.

Peter Maly himself played a very active role in the conception of this book; the 16 pages in the middle were completely left to his discretion. The designer filled them with collages referring to the decades which his design career has spanned. The book thus has an intimate feel to it that makes Peter Maly nearly come to life in its pages.

We asked a great number of friends, colleagues, critics, companions, and manufacturers what Peter Maly is like as a person. Much of the commentary was incorporated into this book, thus enabling an even more highly differentiated view of him. While much of the great wisdom of design may have already served its purpose, or be deemed passé, the »unity of content and form« has survived as quality. Thus, like all books in this series, the foundation of this book is a graphic concept which does justice to its contents.

Acknowledgements:

We would like to thank Karin Görling, Ines Wagemann and Thea Weber for their cooperation, support, and stimulating ideas, as well as all those whose statements on Peter Maly have contributed to the multi-faceted nature of this book.

Alex Buck and Matthias Vogt

Verner Panton

Peter Maly, Perfektionist

»In der Musik von Bach sehe ich eine große Perfektion, die ich in meiner Arbeit zu erreichen suche.« (Peter Maly)

Auf Peter Maly wurde ich vor vielen Jahren aufmerksam, als ich in »Schöner Wohnen« einige seiner äußerst eleganten Möbel- und Einrichtungszeichnungen sah.
Die Gabe, gut zeichnen zu können, birgt für einen Designer Vor- und Nachteile: Zweifellos kann er sich selbst und andere mit einer guten Zeichnung täuschen. Der Vorteil aber liegt in der realistischen Darstellung dessen, was er als Entwurf im Kopf hat – und dies begreife ich als Stärke Peter Malys.
Die Jahre als Innenarchitekt in der »Schöner Wohnen«-Redaktion hatten ohne Frage sowohl für »Schöner Wohnen« als auch für Peter Maly selbst große Bedeutung, fielen sie doch in die interessante Periode der Nachkriegszeit und des Wiederaufbaus. Es war ein Neubeginn.
Sein herausragendes Talent ist es, räumlich denken zu können. Gestaltet er Interieurs oder Ausstellungen, erreicht er eine Perfektion, die ihresgleichen sucht. Und mehr noch: all seinen Projekten liegen gleiche Annahmen und Gedanken zugrunde.

Peter Maly,
perfectionist

»I perceive a great perfection in Bach's music, which I strive for in my own work.« (Peter Maly)

I first became aware of Peter Maly many years ago when I saw some of his highly elegant sketches of furniture and interior decor in the magazine »Schöner Wohnen«. The gift of being able to draw well has both benefits and drawbacks for a designer, one disadvantage being that he can without a doubt fool both himself and others with a good drawing. The advantage, however, is that this talent enables him to portray realistically the sketch he has in his head – and this, as I perceive it, is Peter Maly's forte.

The years which Maly spent as an interior designer on the editorial staff of »Schöner Wohnen« were certainly of great importance for »Schöner Wohnen« as well as for Peter Maly himself. After all, they happened during a most interesting period: the post-war and reconstruction era. It was a new beginning.

Maly's extraordinary talent is the ability to think spatially. When he designs interiors or exhibitions, he achieves a perfection that is without equal. What's more: the same assumptions and thoughts are at the heart of all his projects.

In designing new objects, Peter Maly always strives for longevity. In his eyes, a designer's intention should not merely be to devise concepts only for the present, and to follow the design trends propagated by the media. Rather, he must expand his horizons to include long-term thinking. Responsible research and practicality are equally essential as the basis for the design, as is the highest quality in terms of material,

Peter Maly plädiert beim Entwurf neuer Objekte stets für die Berücksichtigung einer langen Lebensdauer. Es kann seiner Ansicht nach nicht die Absicht eines Gestalters sein, nur für den Augenblick zu konzipieren und von den Medien propagierten Designtrends zu folgen. Vielmehr muß in langen Zeiträumen gedacht werden. Voraussetzung hierfür sind gleichermaßen verantwortungsvolle Forschung und praktische Vernunft als Basis des Entwurfs sowie beste Qualität hinsichtlich Material, Detail und Verarbeitung in der Ausführung. Auch sollte die Form nach Jahren noch ansprechend sein. Für diese Aufgabe eignen sich nach Peter Maly vor allem archetypische Formen und klare, logisch begründete, geometrisch-elementare Grundformen: das ist das bevorzugte Vokabular des Hamburger Designers. »Wir benötigen Rezepte«, lautet seine Devise, »gegen den optischen Verschleiß!«

Gerade in einer Zeit ökologischer und umwelttechnischer Probleme, die sich auch bei der Gestaltung neuer Möbel niederschlagen, fallen die Arbeiten Peter Malys besonders positiv auf.

Viel zu lange haben sich Designer und Produzenten in aller Welt darauf konzentriert, Objekte zu entwickeln, die möglichst schön und günstig sind, nach kurzer Zeit aber regelrecht als Sondermüll entsorgt werden müssen. Diese »Nach-mir-die-Sintflut«-Mentalität können wir uns nicht länger leisten. Peter Maly hat dies längst klar erkannt. Er handelt anders: umsichtig und verantwortungsvoll.

»Ein Stuhl ist für den Hintern gemacht, nicht nur für den Kopf«, hat er einst prägnant formuliert. Der Stuhl ein Kunstwerk? Nicht für Maly: »Trotz aller ästhetischen Ansprüche ist die menschliche Anatomie mit ihren gewissen Rundungen nun einmal nicht wegzudiskutieren. Design muß sich den Zwängen stellen...« Peter Maly hat nicht unrecht. Und doch kann ich ihm nicht uneingeschränkt zustimmen.

Für Stühle, die acht Stunden am Tag ihrem »Besitzer« dienen sollen, mag Malys These durchaus zutreffend sein. Die meisten Stühle jedoch sind häufig verwaist und werden nur kurzfristig zum Sitzen benötigt. Bei solchen Stühlen ist es meiner Meinung nach durchaus denkbar, daß sie auch dem Kopf – genauer gesagt dem Auge – dienen sollen. Mit anderen Worten: sie müssen nicht zwingend bequem und anatomisch richtig konzipiert sein wie beispielsweise ein Bürostuhl, der nicht in erster Linie optisch ansprechend sein muß.

Schaue ich mir nun die von Peter Maly entworfenen Stühle an, kann ich beruhigt feststellen, daß sie – entgegen den Absichten des Gestalters – keineswegs nur »für den Hintern«, sondern auch für den Kopf konzipiert sind. Das eine schließt das andere ja nicht aus.

Auch die Einheit Wohnung muß nach dem Empfinden des Hamburger Designers kein Gesamtkunstwerk im Sinne des Jugendstils sein. Striktes Reglement lehnt er ab. Viel zu diktatorisch und steril erscheinen Maly die Ergebnisse des Postulats, daß alles ein-

Er ist ein funktionsorientierter Ästhet – ein Künstler, ein Ingenieur, ein Designer und ein Innenarchitekt mit besonderen Fähigkeiten zur Visualisierung und Interpretation seiner Ideen.
Jenseits von Memphis, der Postmoderne und anderen Moden entwickelt Peter Maly ernsthaft, nachdenklich, einfallsreich, zielstrebig, stets zum Grundsätzlichen vordringend Möbel und Räume mit unverwechselbarer Ausprägung. Dem Zeitgeist vorauseilend, gelang ihm Bleibendes. Maly-Möbel führen immer zu räumlichen Konsequenzen, die weiterführenden Systeme des Designers sind in ihrer Vollkommenheit Innenarchitektur.
Seine besten »Helfer« sind offensichtlich – neben anderen guten Eigenschaften – Proportionsgefühl, Denkarbeit, Materialfeeling, fachliche Kompetenz sowie umfangreiches Wissen über Fertigungstechniken, Märkte und kulturelle Werte.
Ich war vor langer Zeit einer seiner Lehrer – und ich habe seitdem viel von ihm gelernt.

He is an aesthete with an orientation to functionality – an artist, an engineer, a designer, and an interior designer with a special talent for visualizing and interpreting his own ideas.
Going beyond Memphis, Postmodernism and other trends, Peter Maly develops furniture and rooms with unmistakable character. His approach is serious, contemplative, imaginative, ambitious, and always gets down to the basics. He has created enduring works by staying one step ahead of the *Zeitgeist*. Maly furniture always has spatial consequences, for the designer's furniture systems constitute, in their completeness, interior design.
In addition to other fine qualities, his best »helpers« are obviously a sense of proportion, brainwork, a feeling for the materials, expertise, as well as a comprehensive knowledge of production techniques, markets, and cultural values.
Many years ago I was his teacher – and in the meantime it is I who have learned a great deal from him.

Felix Nitsch

Professor für Innenarchitektur und ehemaliger Dekan an der Fachhochschule Detmold

Professor of interior design and former Dean of the Fachhochschule in Detmold, Germany

detail, and workmanship in its execution. The design should also retain its appeal over the years. According to Peter Maly, archetypal forms and clear, geometric basic elements that are grounded in logic are particularly suitable for thinking in the long-term. This is in fact the preferred vocabulary of the Hamburg designer. As his motto goes, »We need antidotes for optical obsolescence!«

Peter Maly's works are particularly refreshing during a time of ecological and environmental problems which also come to bear in the process of designing new furniture. For far too long, designers and manufacturers throughout the world have focused on developing objects that are as attractive and inexpensive as possible, but which after a short time end up being thrown away and cluttering landfills. We can no longer afford this »après moi le déluge« mentality, something which Peter Maly recognised clearly long ago. He takes a different approach: circumspect and responsible.

»A chair is for the derrière, not for the head«, as he once formulated succinctly. The chair as a work of art? Not for Maly: »Despite all aesthetic claims to the contrary, the human anatomy with its special curvatures is not something one can simply ignore. Design must come to terms with certain constraints ...« Peter Maly is not entirely wrong in what he is saying. And yet I cannot wholeheartedly agree with him.

Maly's hypothesis may thoroughly apply to chairs which are intended to serve their »sitter« for eight hours a day. However, the majority of chairs are often »abandoned«, in that they are used only briefly for sitting. As I see it, it is quite conceivable for them to serve the head as well, or more specifically, the eye. In other words, they do not necessarily have to be as comfortable and anatomically correct as an office chair, for example, which in turn does not have to be visually appealing.

When I look at the chairs designed by Peter Maly, I find, reassured, that despite the designer's intentions, they are in no way merely designed »for the derrière«, but also for the head. After all, the one does not necessarily preclude the other.

Nor, as Maly sees it, does »living space« as a unit have to be a gesamtkunstwerk à la art nouveau. He rejects the application of strict rules and guidelines. The results of the postulate that everything has to follow uniform standards are much too dictatorial and sterile for Maly's taste. »A living space must be able to grow«, he explains. »It should be able to add annual rings like a tree, just as a well-designed piece of furniture may also age gracefully.«

Once again, I cannot completely agree with Peter Maly. If someone needs to live in a gesamtkunstwerk that has been styled from top to bottom in order to really feel »at home« there, then he should do so. However, he should do so only because he himself wishes to do so, and not because it has been thrust upon him by design proselytes and trend-setters.

heitlichen Maßgaben zu folgen habe. »Eine Wohnung muß wachsen«, begründet der Gestalter, »sie darf Jahresringe ansetzen, ebenso wie auch ein gut gestaltetes Möbelstück altern darf.«

Auch in diesem Punkt bin ich mit Peter Maly nicht ganz einer Meinung. Hat jemand das Bedürfnis, in einem durchgestylten Gesamtkunstwerk zu wohnen, und fühlt er sich tatsächlich darin wohl, soll er es auch tun. Aber nur, weil er selbst es möchte und nicht, weil es ihm von Design-Predigern und Trendsettern aufgezwungen wird.

Peter Maly ist ein ernsthafter und ernstzunehmender Designer. Seine Glaubwürdigkeit wird dadurch unterstrichen, daß er seine Meinung und Überzeugung ohne Kompromisse umsetzt. Für ihn, den naturverbundenen Menschen, haben Umwelt und ökologisch sinnvolles Handeln einen sehr hohen Stellenwert.

Der gelernte Tischler duldet keine Halbheiten, wenn es um Materialqualität und Ausführung geht. Als multitalentierter Künstler spielt er überzeugend mit Farben und Formen, Konstruktionen und Kontrasten. Seine Objekte sind nie schrill oder aufdringlich, vielmehr realitätsbezogen und anwendbar. Und nicht zuletzt: aufrichtig. Ganz wie der Designer selbst.

Wer in den vergangenen 35 Jahren engagiert und leidenschaftlich seinen Beruf gelebt hat, weiß, was es bedeutet, über einen solch langen Zeitraum eine Person im Gedächtnis zu bewahren.
Peter Maly und ich waren Studienkollegen und pflegten eine eher distanzierte Freundschaft; man beobachtete sich.
Die Präsenz einer Person über jahrzehntelange Distanz begründet sich mehrschichtig. Das herausragende Element meines Verhältnisses zu Peter Maly aber ist die Bestätigung seiner von Anfang an erkennbaren Begabung, seines hohen Qualitätsanspruchs und der großen Beharrlichkeit des Hamburger Designers. Es freut mich sehr, daß ich mein frühes Urteil nie revidieren mußte.
Peter Maly gehört ohne Zweifel zu den führenden deutschen Gestaltern, die auf hohem Niveau Serienprodukte entwerfen, welche dank ihrer großen Auflagen einer breiten Käuferschicht zugute kommen. Diese Leistung verdient besondere Wertschätzung, zumal sie so wenig in unsere Zeit passen will.

Anyone who has been passionately involved in their profession over the past 35 years knows what it means to keep someone in mind for such a long period.
Peter Maly and I were fellow students, and maintained a distant friendship; we observed each other, so to speak.
A person's presence acquires multiple layers over the course of so many years. However, the most outstanding element of my relationship to Peter Maly has always been the consistent affirmation of his talent which was apparent from the very beginning, his high quality standard and his intense perseverance. I am pleased to know that I never needed to revise my earlier assessment of him.
Without a doubt, Peter Maly belongs to the group of leading German designers who design high-quality product lines that can be manufactured in large quantities to the benefit of a broad spectrum of consumers. This achievement deserves special praise, especially as it occurs so rarely in our times …

Hans-Dieter Nieländer

Architekt SWB in Zürich

Architect SWB in Zurich, Switzerland

Peter Maly is a serious designer who is also to be taken seriously. His credibility is underscored by the fact that he turns his opinion and convictions into action without compromising them. As a person with deep roots in nature, he assigns a very high priority to the environment and ecologically sensible behaviour.
Maly, who was trained in cabinetmaking, will accept nothing short of perfection when it comes to the quality of the material and the execution of the design. A multi-talented artist, he plays with colours and forms, constructions and contrasts, in a convincing way. His objects are neither shrill nor obtrusive; rather, they are always attractive and useful. And not least of all, upstanding – just like the designer himself.

Josef Kremerskothen

Wegstücke

Erste Entwurfserlebnisse hatte der Junge im Schneideratelier des Vaters, das dieser mit zwei Dutzend Gesellen in Trautenau, auf der böhmischen Seite des Riesengebirges, betrieb. Den Sohn fesselte es, mit welchem Geschick der Vater den Tuchballen abrollte, mit großen Scheren im Nu den Zuschnitt zauberte, mit Reihfäden heften ließ, bei Anproben korrigierte und schließlich dem Kunden wenige Tage später das wertvolle Stück übergab.

Dann die Flucht. Fortan lebte Peter Maly mit Mutter, Bruder und Verwandten in einem verwunschenen Waldwinkel. Diese Idylle machte das armselige Holzhaus erträglicher. Hab und Gut hatten die Malys mit der böhmischen Heimat verloren.

Die Mutter rackerte für das Wohl der Familie, da der Vater aus dem Krieg nicht heimgekehrt war.

»Wir gehörten im hessischen Dorf«, so Peter Maly, »zu den mißtrauisch beäugten Flüchtlingen, waren Habenichtse, Außenseiter. Freunde habe ich dort kaum gefunden. Aber ich konnte zeichnen und malen. Damit verbrachte ich meine Zeit am liebsten – und mit meinem ›Stabilo‹-Baukasten. Im Wald baute ich geheime Hütten – Sehnsüchte nach einem eigenen Gehäuse. Dort in der Abgeschiedenheit habe ich gelernt, mich auf mich selbst zu besinnen, nicht aufzugeben, eigene Stärken konzentriert und auch spielerisch zu vertiefen. Vielleicht haben die Entbehrungen meine Traumwelten ausgeweitet.«

Stages along the way

The boy had his first experiences with design in his father's tailor's shop in Trautenau, on the Bohemian side of the Riesengebirge. The son was spellbound by the skill with which his father unrolled the bale of cloth, created patterns in the twinkling of an eye with his large scissors, basted the material with thread, made adjustments during the fitting, and then handed over a valuable article to the customer only a few days later. Then the family was forced to flee. From that point on, Peter Maly lived with his mother, brother and relatives in an enchanted corner of the woods. These idyllic surroundings made it easier to bear the wretched wood house in which they were living. The Malys had lost everything they owned along with their Bohemian hometown. As the father did not come home from the war, Peter's mother worked her fingers to the bone to earn a living for the family.

»In that Hessian village«, says Peter Maly, »we were refugees who were regarded with mistrust; we were have-nots, outsiders. I hardly made any friends there. But I could paint and draw. I was happiest spending my time this way – and with my ›Stabilo‹ building set. I built secret huts in the forest – a longing for my own cocoon. There, in these secluded surroundings, I learned to reflect upon myself, not to give up, to develop my own strengths through concentration and also playfulness. Perhaps having to do without helped to expand my dreamworld.«

During his training in cabinetmaking – even here, he was eyed with suspicion (»Peter wants to study, you know«) – he suddenly received unaccustomed attention for a kidney-shaped table he had designed himself (Nierentisch, 1955), which was immediately reproduced in a mini-series for friends. »In building this table, I refined my craftsman's skills; through daily toil I learned to tame the recalcitrant wood, test its hardness and create lovely effects with the surface.«

Das »Froschaugenobjektiv« verlieh diesem Titel galaktische Visionen. Entschlossenen Individualismus zeigt dieser futuristische Raum: Rundungen statt rechten Winkeln, abgehängte Deckenteile, nach Maß gebaute Sitzlandschaft, neuartige Einzelstücke.

The »frog's eye lens« opened up a galactic perspective on this title. This futuristic room bears testimony to resolute individualism: curvatures instead of right angles, suspended ceiling segments, seating landscapes made to measure, innovative unique furniture pieces.

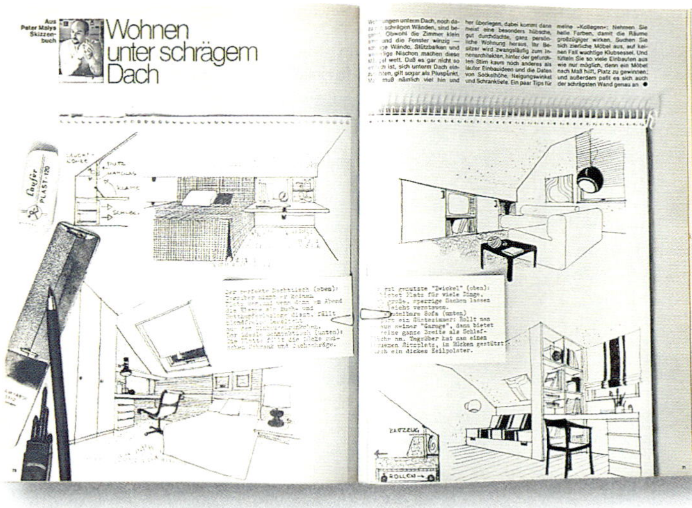

Linke Seite
»Peter Malys Skizzenbuch«, jahrelange Themenserie. Hier Wohnecken unterm schrägen Dach. Ein Dutzend Beispiele, wie man ungenutzte Raumreserven pfiffig verwenden kann. Solche Details lassen sich mit Zeichnungen präzise erklären.

Rechte Seite oben
Materialcollagen ergänzten Raumperspektiven, zeigten Farben und Strukturen unverfälscht, erleichterten die Suche beim Händler. Weder Schatten noch Überstrahlungen, die in räumlichen Farbfotos nicht zu vermeiden sind, veränderten den Eindruck.

Rechte Seite unten
Auf die Farbmengen kam es hier an. Didaktische Lehrstücke lassen sich nur mit gleichem Bildausschnitt und gleichen Einrichtungsstücken begreifbar machen. Apfelgrün, Twiggy-Zeiten.

On the left
»Peter Maly's Sketchbook«, a series focusing on many different themes over the years. Here, the topic is living-rooms with slanted ceilings. A dozen examples of how the unused extra space can be cleverly put to use. The illustrations serve as precise explanations of the details.

Above right
Material collages supplemented room perspectives, illustrated the colours and structures as they would appear in the actual setting, and facilitated the search for the right piece of furniture in the furniture store. The impression was altered neither by shadow nor overly-bright lighting – unavoidable in colour photos of a room.

Below right
Here, it was a question of how much paint to use. Only with the same details and the same furnishings could this lesson be truly educational. Apple green, the Twiggy era.

In der Tischlerlehre – auch dort nörglerisch belauert (»Der Peter will ja studieren«) – plötzlich ungewohnte Aufmerksamkeit durch einen selbstentworfenen Nierentisch (1955), der sogleich in Mini-Serie für Freunde nachgebaut wurde. »Hier habe ich mein handwerkliches Geschick verfeinert, in alltäglicher Mühe gelernt, widerspenstiges Holz zu zähmen, seine Festigkeit zu erproben und den Oberflächen schöne Wirkungen zu verleihen.«

Nach der Gesellenprüfung begann mit dem Studium an der Fachhochschule für Innenarchitektur und Möbeldesign in Detmold die Zeit erster vorbehaltloser Anerkennung. Peter Maly war ein begabter Zeichner. Ihm gelangen plausiblere und auch schönere Darstellungen. Hier fragte man ihn um Rat, seine Meinung war anderen wichtig. Mit einer bescheidenen Studentenbude endlich die eigenen vier Wände, die obendrein freiheitliche Gefühle stärkten. »Im Studium machte ich meine ersten Erfahrungen in Aquarellen, Gouachen und Federzeichnungen mit selbstentworfenen Räumen, mit fein abgestimmten Proportionen, mit harmonischen Farbklängen und mit Materialkompositionen. In meinen Projekten erlebte ich Gegenstände im Ensemble nach ureigenen Vorstellungen.«

Seine Diplomarbeit (Thema: »Liebenswertes Wohnen«) collagierte Peter Maly mit Fotos aus der eben neu erschienenen Zeitschrift »Schöner Wohnen«. In dieser Publikumszeitschrift fand er authentische Bilder, die vom neuen Wohnen aus aller Welt berichteten. Er verschlang, was er in wenigen Büchern und Blättern an interessanten Beispielen finden konnte.

Ein Freund zeigte ihm die längst vergilbte Anzeige, in der »Schöner Wohnen« Innenarchitekten suchte. Peter Maly fand die Bewerbung anmaßend, die er nach freundschaftlichem Drängen schließlich doch auf die Post gab. In Hamburg stand er in der Burchardstraße beklommen vor dem ehrwürdigen Sprinkenhof; am liebsten hätte er kehrtgemacht.

After passing his journeyman's examinations, he began studies at the Detmold Fachhochschule. It was then that his skills first received unreserved recognition. Peter Maly was the best draftsman. People asked him for advice; his opinion was valued by others. »During my studies, I worked for the first time in water-colours, gouache and pen-and-ink drawings, with self-designed spaces, with finely matched proportions, with harmonic hues and with material compositions. In my projects, I experienced objects as parts of an ensemble based on my very own ideas.«
Peter Maly supplemented his final-year project (subject: »attractive living«) with collages of photographs from the new popular magazine »Schöner Wohnen«, where he found authentic pictures portraying new lifestyles from around the world. He devoured everything he could find in the way of interesting examples in the few books and leaflets available.

Ich habe Peter Maly einmal als collagierenden Minimalisten bezeichnet.
Er besitzt die große Fähigkeit, Formen, die im Trend liegen, aufzugreifen und zu abstrahieren. Denn abstrakte oder minimale Formen sind ihm ein wesentliches Bedürfnis. Sie in seinen Produkten zu collagieren, ist ein weiteres Merkmal seiner Arbeit.
Beides zusammen garantiert ihm oft seine Akzeptanz beim Publikum.

I once described Peter Maly as a minimalist who makes collages.
He possesses the great talent of being able to extract forms promulgated by the current trend and create abstractions from them. For abstract or minimal forms are an essential need for Maly. Forming collages with them in his products is an additional trademark of his work.
Combining these two processes in his work often guarantees him public acceptance.

Dr. Gudrun Scholz

Professorin für Designtheorie an der Fachhochschule in Bielefeld

Professor of design theory at the Fachhochschule Bielefeld, Germany

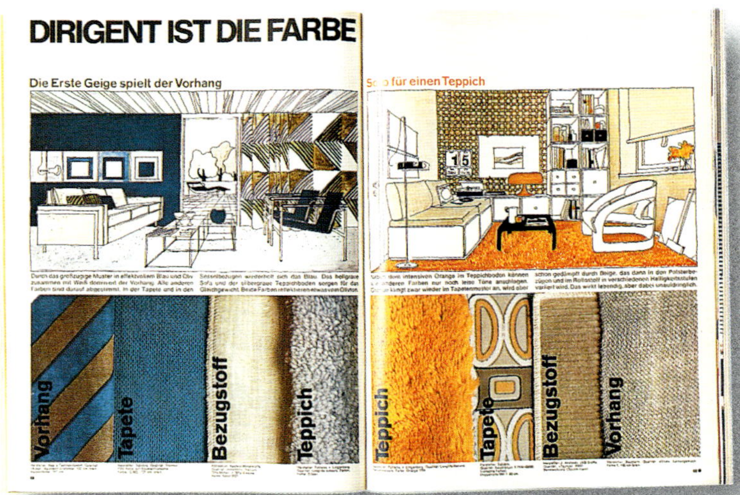

Die Redaktion hatte Bewerbungen bergeweise erhalten, die Entscheidung jedoch fiel erst nach dem Gespräch mit Peter Maly. »Der brachte als einziger druckfertige Zeichnungen, glühte vor Begeisterung – ein vitaler Querdenker, der mit uns anderen heilmachen wollte, was kaputtgegangen war. Wenn es darauf ankam, wußte er auf verblüffende Art Türen zu öffnen«, so Hasso G. Stachow, »Schöner Wohnens« erster Chefredakteur.

Der stille Bursche aus der Provinz hatte Mühe in der ungewohnt urbanen Welt und auch mit den ungenierten, manchmal naßforschen Journalisten, denen zunächst seine – vielleicht zu große – Hochachtung galt. Die Nachrichtenleute reagierten auf Ereignisse von draußen, Stoff für ihre Geschichten – so hatten sie es immer gemacht. Der Innenarchitekt wurde anfangs als trefflicher Zeichner und Nützling für unliebige Fachfragen in Anspruch genommen.

Mit einem Mal war da aber einer, der Nachrichten mit seinen Entwürfen in allen Einzelheiten selbst schuf – bestaunter und beargwöhnter Themenerfinder zugleich. Ein Sakrileg wider das journalistische Ethos? Da wurde zäh und leidenschaftlich gestritten, bis aus Einzelkämpfern ein Gespann zusammengewachsen war. Bald entstanden in improvisierten Ateliers Raumkonzepte im Maßstab eins zu eins: wirkliche Räume mit allem Drum und Dran, richtig zum Anfassen – unfaßbar. Besserwisserische Kritiker wandelten sich in stolze Miterfinder, denn begeisterte Leserbriefe und eine rasch steigende Auflage waren Lohn für das engagierte Team.

Es war ein Beginn fast ohne Vor-BILDER. Wohnzeitschriften für das breite Publikum gab es nach dem Krieg nicht –»Schöner Wohnen« wurde Vorreiter. Was die Redaktion aus dem Ausland – zunächst spärlich – erreichte, konnte wenige Anregungen geben. Die Franzosen waren pompös, die Engländer verschroben, die Amerikaner zu großräumig. Brauchbarer waren skandinavische Zeitschriften, aus denen mancherlei Anregungen gewonnen wurden. Während Möbel und Wohnutensilien aus anderen Ländern hier noch keinen Markt hatten, waren Dänen, Schweden und Finnen längst vertreten. Die Teakwelle rollte mächtig und schwappte in die guten Stuben.

Faszinierender waren für die Redaktion aber die Schöpfungen der großen Architekten: Alvar Aalto, Arne Jacobsen, Verner Panton, Poul Kjaerholm, Hans Wegener, Poul Henningsen und viele andere. Sie formten ihre wunderbaren Gebilde vor allem aus Holz, das uns Menschen seit alters nahe ist. Poul Henningsen damals: »Wir sind hier im Lande sicher innig mit dem dänischen Holz verlobt – das wir aus den fernsten Gegenden der Welt importieren, um es später schön geputzt wieder zurückzuschicken.«

In diesen ersten Jahren waren die Messen in Kopenhagen das Eldorado: Möbel, Leuchten, Stoffe, Geschirr, Teppiche – alles was zur Wohnung gehörte, fand sich hier; und alles konnte harmonisch im Ensemble auftreten. Peter Maly war hingerissen, wäre am liebsten in Kopenhagen geblieben und hatte Mühe, in seinen Entwürfen Gegenstände aus anderen Regionen zuzulassen.

Wir kennen uns seit Anfang der sechziger Jahre, gleichaltrig, als Innenarchitekten an verschiedenen Zeitschriften im selben Verlag engagiert. Fasziniert von den Möglichkeiten, die das damals neue Medium bot.
Peter Maly entwickelte für »Schöner Wohnen« eine unverwechselbare Darstellung. Später für »Interlübke« eine gestalterische Qualität, die das Unternehmen ohne ihn nicht wieder erreichte.
Uns verbindet bis heute ein freundschaftlicher Respekt. Bemerkenswert für mich sind die Perfektion all seiner Ergebnisse, der elegante Umgang mit Material, Form und Farbe, immer mit Stilgefühl und immer auf der Höhe der Zeit.

We have known each other since the Sixties. At that time, we were working for different magazines in the same publishing company – two interior designers of the same age. Fascinated by the possibilities offered by what was then a new medium.
Peter Maly developed an unmistakable presentation style for »Schöner Wohnen«. And later, for »Interlübke«, a quality of design that the company never attained again after he left.
Today, we are linked by a friendly respect. I find the perfection of all of his products remarkable – his elegant handling of the material, form and colour, always with a feeling for style, and always at the cutting edge.

Rolf Heide

Designer und Innenarchitekt in Hamburg

Designer and interior designer in Hamburg, Germany

FARBREZEPTE für die Wohnung

Die Grundfarben	Nachbarfarbe	Ton-in-Ton	Komplementärfarbe

Farbigkeit als Ausdruck des jungen Wohnens

Eine reizvolle Dekoration nach Art der Op Art

Von oben nach unten

Farbrezepte für Wohnräume. Lesern mit Vorlieben für eine der vier Farben – Gelb, Rot, Blau und Grün – wurden Farbverteilungen für ihre Lieblingsfarbe in identischen Raumperspektiven erklärt.

Lieblingsfarbe Rot in weißem Raum: Erregend, dynamisch, aktiv. Besonders junge Leser ließen sich von den Pop-Farben Ende der sechziger Jahre hinreißen und erhielten Rat von den »Schöner Wohnen«-Architekten.

Der Maler Victor Vasarély, die »Op-Art« faszinierten auch Peter Maly. Das Streifendessin ließ die Türen der Schrankwand – rechts und links des Bettes – »verschwinden«. Mit solchen Experimenten sprengte Maly zuweilen den Rahmen nachvollziehbarer Wohnanregungen.

From top to bottom

Colour solutions for living spaces. Readers with a predilection for the four specific colours – yellow, red, blue and green – were presented with ways of using their favourite colours based on identical room perspectives.

Red in a white room: stimulating, dynamic, active. Young readers in particular were enthralled by pop art colours at the end of the Sixties, and received advice from the designers at »Schöner Wohnen«.

Peter Maly was fascinated by painter Victor Vasarély and Op Art. The striped design renders the cupboard doors – to the left and right of the bed – »invisible«. With these experiments, Peter Maly sometimes exploded the habitual in interior design concepts.

Im August 1961 widmete die dänische Zeitschrift »mobilia« Verner Panton und Jean Tinguely – zwei überragenden Formenschöpfern jener Jahre – 113 Seiten. Dieses Ideenfeuerwerk, diese hinreißende Dokumentation war bald vergriffen, wurde zum heute kaum noch auffindbaren publizistischen Sammeljuwel.
Eine anschwellende Flut von Leserbriefen hielt die Redaktion auf Trab und sorgte für Tagesnähe. Hilferufe waren das, die gewissenhaften Rat der Redaktion herausforderten. Um 1960 wohnten die meisten Menschen beengt. Die ständig wiederkehrende Frage lautete daher: Wie läßt sich aus der Not eine Tugend machen?
Kleine Räume durften nicht eng, Improvisation durfte nicht primitiv, sondern mußte pfiffig wirken. Ergänzungen sollten harmonisch und nicht beliebig ausfallen. Das

Es ist zwiespältig für mich, über Peter Maly zu schreiben, denn für »Möbel perdu« als Protagonist der Neuen Deutschen Designbewegung in den achtziger Jahren war er einer der Antipoden: postmodern, mainstream und vor allem ein Hardcore-Industrie-Designer.
Beide haben wir uns in den neunziger Jahren respektierend gewandelt und kooperieren mittlerweile. Seine elegante Stuhlkreation für Thonet mit dem Gurtgeflecht begeistert mich geradezu: ein moderner Klassiker! Ich bin auf seine weiteren Entwicklungen gespannt, denn er versteht die Kunst, serielle Systembauweisen mit innovativem Handwerk zu kultivieren.
Heute empfinde ich Peter Maly als einen besonnenen, mehr aus dem Hintergrund agierenden Designer mit gemütvollem, erdverbundenem »Rübezahlcharme«. Seine Arbeit strahlt, auch durch die Wahl der Materialien, eine sinnlich handwerkliche Solidität aus, in einer Formensprache, die stark vom Bauhaus inspiriert und dem Quadrat verschrieben ist. Dies bedeutet für ihn Harmonie und Ausgewogenheit, denn, so Maly, »ein Quadrat hat keine Richtung«.

I have mixed feelings when writing about Peter Maly, for as a protagonist of the New German Design movement in the Eighties, he was one of the antipodes to »Möbel perdu«: post-modern, mainstream, and above all a hard-core industrial designer.
Both of us have changed during the Nineties, and in the meantime have learned to respect and co-operate with one another. I am an absolute fan of the elegant chair he created for Thonet, with its woven strips – a modern classic! I am looking forward to his upcoming creations, for he understands the art of cultivating serial system constructions through innovative craftsmanship.
Today, Peter Maly is for me a thoughtful designer who tends to work behind the scenes, with a warm-hearted, earthbound, elfish charm. His work radiates the solidity of a craftsman's touch, which makes itself instantly felt, in part owing to the choice of material. His formal idiom has been inspired to a great extent by the Bauhaus, and is devoted to the square. For him, the square means harmony and equilibrium, and in Maly's own words, »A square takes no direction«.

Claudia Schneider-Esleben

Architektin, Designerin und Galeristin in Hamburg

Architect, designer and gallery owner in Hamburg, Germany

It was a friend who showed him the employment ad in which »Schöner Wohnen« was looking for interior designers. Maly sent off his application reluctantly, at the urging of his friend. The editorial staff had received a veritable mountain of applications, but made a decision only after the interview with Peter Maly. »He was the only one to bring drawings ready for printing, he was glowing with enthusiasm – a vital and original thinker, who wanted to work with the rest of us to get things back on the right track. In a pinch, he knew how to open doors if he had to, and in a most astounding manner«, reminisces Hasso G. Stachow, the first editor-in-chief of »Schöner Wohnen«. The quiet lad from the country had trouble with the unfamiliar urban world, as well as with the unabashed, sometimes brash journalists, whom he at first held in perhaps too high esteem. In the beginning, the interior designer was used for his excellent draftsman's skills and as a sounding board for difficult specialist questions.

Then, all of a sudden, here was someone who was making news himself with his designs – at once the subject of admiration and suspicion. Was this a sacrilege to the ethos of journalism? Heated debate on this point gave way to teamwork, as know-it-all critics turned into proud co-inventors. Soon afterwards, concepts for rooms emerged in improvised studios on a scale of one-to-one – real rooms with all the trimmings, which one could reach out and touch.

It was a beginning almost devoid of role models. Interior design magazines for the general public did not exist after the war. »Schöner Wohnen« was a pioneer. The achievements of foreign publications in the field – at first sparse – were in any case not particularly inspiring. The French were pompous, the English were eccentric, the Americans too expansive. Journals from Scandinavia were more useful. While there was no market in Germany as yet for furniture from other countries, the Danes, Swedes and Finns had been represented for ages.

Linke Seite
Großer Atelieraufbau, in welchem die Kulissenwelt den Lesern nicht verschwiegen wurde. Solche Breitwand-Demonstrationen sollten überraschende Raumübergänge erklären, den Einklang von Farben, Materialien und Strukturen, den Lichtrhythmus, Veränderung der Raumproportionen mit Hilfe von Podesten oder abgehängten Deckensegmenten.

Rechte Seite
Kunststoffe – die Zukunft hatte für die Elite der Designer und Architekten begonnen. Solche »Skulpturen« hatte es bisher nicht gegeben. Eine Maschine konnte in Sekundenschnelle das Werk vollenden, wo bislang ein Möbel aus Holz erst nach vielen Arbeitsgängen die fertige Form bekommen hatte.

On the left
Large studio construction in which the wings were not concealed from the readers. These panoramic demonstrations were intended to explain startling spatial transitions, the harmony of colours, materials and structures, and the rhythm of the lights; changes in the proportions of the room were elucidated with the help of platforms, or suspended ceiling segments.

On the right
Plastics – the future had begun for the design and architectural elite. These »sculptures« had never existed before. The work could be completed in a matter of seconds by a machine, whereas wooden furniture only acquired its finished form after countless processing stages.

Licht mußte stimmungsvolle Wohnatmosphäre erzeugen. Die Zimmer durften nicht vollgestopft erscheinen, die Farben mußten mit Materialien fein abgestimmt sein. Niedrige Räume sollten nicht bedrückend, hohe nicht aus dem Lot geraten wirken. Die Leser wollten es nicht altväterlich, eher zeitgemäß, wünschten statt modischer Kühle Wärme und Behaglichkeit.
Mit einer mächtig wachsenden Möbelindustrie kamen auch zunehmend Anzeigen ins Heft. So nahm der Heftumfang zu. Immer mehr Produktionen erforderten professionellere Ateliers, in denen mehrere Entwürfe gleichzeitig aufgebaut werden konnten. Die Redaktion warb weitere Mitarbeiter an. Unter der Stabführung von Peter Maly wirkten zeitweilig sieben festangestellte und vier freie Innenarchitekten.
Die Redaktion bot begehrte Arbeitsmöglichkeiten. Innerhalb eines Jahres konnte ein Architekt in diesen verkürzten Bauprozessen mehr Raum-, Licht-, Material-, Farb- und Ensembleerfahrungen machen, als das in einem ganzen Architektenleben sonst möglich gewesen wäre: alles authentisch, im Maßstab eins zu eins. Wenn die Farben bei Kunstlicht nicht mehr »stimmten«, wurden Wände und Decken neu gestrichen, Accessoires ausgetauscht, das Licht geändert, Möbel gewechselt. Im Team mit den Journalisten wurden Entwerfer auch Publizisten. Schließlich wollten die Entwurfsabsichten anonymen Lesern plausibel vermittelt werden. Schon in den sechziger Jahren arbeitete man für beinahe drei Millionen Leser pro Monat, weil jedes Heft damals zehn bis fünfzehn Mitleser hatte.
Da im Atelier zuweilen Platzmangel herrschte, halfen perspektivische Zeichnungen mit Detailschnitten. So entstand die Serie »Peter Malys Skizzenbuch« (Seite 18). In jeweils zwei Seiten steckten über ein Dutzend Anregungen für das Wohnen, meist Extrakte aus Leserfragen. Gezeichnete Beispiele mit Material- und Farbcollagen erleichterten dem Betrachter das Nachmachen (Seite 19).

Von links nach rechts
Kunststoffe konnten in allen gewünschten Farben gegossen werden. Meist war das Material durchgefärbt – Beschädigungen fielen kaum noch auf.

Die Hauptzeile – so provokativ wie das plakative Titelbild – weckte den Widerspruch vieler Leser, begeisterte die Avantgardisten, löste eine Flut von Leserbriefen aus und setzte Gespräche in Gang.

From left to right
Plastic could be moulded into whatever form was desired. The material was usually dyed through, which meant that surface damage was hardly noticeable.

The headline – as provocative as the graphic title picture – elicited protests from many readers, inspired the avant gardes, triggered a flood of readers' letters, and stimulated conversation.

However, even more fascinating for the budding publication were the oeuvres of the great architects: Alvar Aalto, Arne Jacobson, Verner Panton, Poul Kjaerholm, Hans Wegener, Poul Henningsen, and many others. They created their wonderful objects in particular out of wood, a substance with which humans have felt an affinity for aeons. As Poul Henningsen put it at the time, »Here in this country, we definitely have an intimate relationship with Danish wood at some internal level – we import it from the most remote areas of the world, only to send it back again nicely polished.«

During these early years, the trade fairs in Copenhagen were the Eldorado of the branch: furniture, luminaires, fabrics, tableware, carpets – everything which belongs in a home could be found here, and everything could be mixed and matched as part of a harmonic ensemble. Peter Maly was enthralled. He would have most preferred to stay in Copenhagen, and had difficulties accepting objects from other regions in his designs.

In August 1961, the Danish magazine »mobilia« devoted 113 pages to Verner Panton and Jean Tinguely. This enchanting documentation was soon out of print; in the meantime, it has become a rare collector's item, seldom to be found.

A growing flood of letters from readers asking for advice kept the editors busy, and also in touch with day-to-day concerns. Around 1960, most people were living in cramped conditions. Thus, the question which was constantly being asked was: how to make a virtue of necessity?

Small rooms could not look cramped, improvisation could not seem primitive – they had to appear lively. Supplementations were supposed to create harmony and not seem arbitrary. Light had to create atmosphere. Rooms could not look as if they had been stuffed with furniture; colours had to be finely matched with materials. Low-ceilinged rooms could not appear oppressive, while those with high ceilings were not supposed to look as if something were out of kilter. Readers did not want old-fashioned patriarchal styles, but instead preferred a contemporary look, and in place of fashionable coolness, they desired warmth and coziness.

The rapidly expanding furniture industry led to more advertisements being included in the magazine, which began to increase in scope. A growing number of photo sessions required studios that were more professional, in which a greater number of designs could be developed at once. The editorial staff added new members.

At times Peter Maly was in charge of as many as seven full-time employees and four free-lance interior designers.

The magazine offered coveted job opportunities. By working with these abbreviated construction processes, a designer could gain more experience in terms of space, light, material, colour and ensemble concepts in one year than would have otherwise been possible in an entire lifetime. And all this completely authentic, on a scale of one-to-one. Working together with journalists, designers became journalists as well. Even in the Sixties, this team was working for nearly three million readers every month, for in those days every copy was shared by 10 to 15 readers.

In der »Design Encyclopedia« findet sich unter »M«: »Maly, Peter (* 1936), German designer, clients included COR, Germany, for whom he designed 1984 ›Zyklus‹ chair, and Ligne Roset, France, for whom he designed numerous pieces of furniture, including 1983 ›Maly bed‹.«

Das war´s. Ein kurzer Eintrag. Mehr nicht. Peter Maly wird in der internationalen Design-Literatur knapp gehalten.

In der Szene der Professionellen steht sein Name hingegen für einen der meistbeschäftigten Möbelgestalter, als Entwerferpersönlichkeit, die schon mit Entwürfen der sechziger Jahre »moderne Klassiker« schuf.

Seinen Arbeiten haftet nichts Spektakuläres an. Peter Maly ist kein »Künstler«. Der Innenarchitekt Peter Maly ist gelernter Tischler. Und das galt in den fünfziger Jahren als solide Basis für die Konstruktion von Möbeln.

Als ich Mitte der siebziger Jahre den Entwerfer meines Lieblingsmöbels, des Pinienschranks »Quadrat«, auf dem Messestand kennenlernte, war ich beeindruckt. Während nebenan die »Avantgardisten« unwirkliche Wohnwelten zelebrierten, faszinierte der junge Maly als Industrie-Designer, der mit durchdachten Möbelprogrammen und ausgefeilten Einrichtungssystemen Strukturen künftigen Wohnens entwickelte – realistische Muster möglicher Wohnwelten.

If you look in the »Design Encyclopaedia« under »M«, you will find the following: »Maly, Peter (*1936), German designer, clients included COR, Germany, for whom he designed 1984 ›Zyklus‹ chair, and Ligne Roset, France, for whom he designed numerous pieces of furniture, including 1983 ›Maly bed‹.«

Die umfangreichen eigenen Produktionen machten es erforderlich, daß neben exzellenten Designerstücken – oft aus dem Ausland – auch »stinknormale« Massenmöbel eingeplant werden mußten. Das war wenig anregend, aber realitätsnah, wollten die Leser doch zu eigenen Banalitäten Ergänzungen finden. Diese Herausforderung führte so manches Mal zu interessanten, spannungsvollen Lösungen.
Peter Maly inspirierte der Mangel auf dem deutschen Markt zu einer Serie kleinerer, praktischer, beweglicher Ergänzungsmöbel in Weiß, Orange, Pink und Oliv. Kultfarben jener Zeit. Tecta baute später diese Stücke, die eine ungewöhnliche publizistische Resonanz – sogar in »abitare« – fanden (Seite 26 und Seite 29). Dieser Erfolg hatte bei Peter Maly Designerinstinkte geweckt. Wenig später entwickelte er für COR den Schalensessel »Trinom«, als Element in Reihe oder als Einzelsessel geeignet (Seite 31).
Große Schwierigkeiten hatten die Leser auch mit Farben. Aus der Mode drangen leuchtende Op- und Popfarben in die Wohnung: Pink, Orange, Saharagelb und Apfelgrün beherrschten die Farbskalen der Designer. Farbbeispiele gaben die Innenarchitekten ohnehin in allen Entwürfen, nun aber wurde Farbe im Heft zum zentralen Medium.
Peter Maly ging das Thema didaktisch an (Seite 21 oben), indem er Phänomene wie »Grundfarben«, »Nachbarfarben«, »Komplementärfarben« und »Ton-in-Ton-Farben« mit 16 Beispielen in exakt gleicher Raummöblierung zum Einstieg ins Thema zeich-

And that's it. A short entry, nothing more. Peter Maly is kept brief in international design literature.
By contrast, in the professional scene, his name stands for the most active furniture designer, a prominent designer who has already created »modern classics« with his designs from the Sixties.
There is absolutely no trace of the spectacular in his works. Peter Maly is not an »artist«. Interior designer Peter Maly was trained as a cabinetmaker, which in the Fifties was a solid basis for building furniture. When I met the designer of my favourite piece of furniture, the pine cabinet »Quadrat«, at the trade fair exhibit in the mid-Seventies, I was impressed. While the »avant-gardes« were celebrating unreal home interiors at the stand next to us, young Maly proved fascinating as an industrial designer who developed structures for homes of the future with his well thought-out furniture programs and flawless furnishing systems – realistic models for possible interiors.

Karlheinz Krug

Chefredakteur und Mitherausgeber der Zeitschrift »form« in Düsseldorf

Editor-in-chief and co-publisher of »form« magazine in Düsseldorf, Germany

Because the studio sometimes lacked space, perspective drawings and detailed sections were of great help. This is how the »Peter Maly's Sketchbook« series came into being (page 18). Two pages of each issue contained over a dozen ideas for interiors, mostly referring to readers' questions. Exemplary sketches with collages of material and colours made it easy for the observer to imitate (page 19 above).

The magazine's extensive own production made it necessary to plan in »plain old« mass-produced furniture in addition to excellent designer pieces, many of which originated outside Germany. While not particularly stimulating, this approach was realistic, for readers wanted ways to supplement their own banality. The challenge sometimes led to exciting solutions.

The dearth in the German market led Peter Maly to produce a series of smaller, more practical, more mobile supplementary furniture in white, orange, pink and olive – the cult colours of the time. These pieces were later produced by Tecta, and they found an unusual degree of journalistic resonance, even in »abitare« (page 26 and 29). Soon afterward, he developed the »Trinom« shell-type chair for COR, suitable as an element in a series or as an individual chair (page 31).

Readers also had great difficulties with colours. Glowing op and pop art colours from the fashion scene barged into people's homes: pink, orange, Sahara yellow and apple green dominated designers' colour scales. Interior designers had always supplied colour samples with their designs, but now colour was becoming a central medium in the magazine. Peter Maly tackled the theme didactically (page 21 above) by illustrating phenomena such as »primary colours«, »adjacent colours«, »complementary colours« and »matching colours«. He used 16 examples with exactly the same set of furniture as a way of approaching this subject. Enthusiasm often induced laypeople to make excessive use of their paint buckets. The examples shown on page 19 below were attempts to give readers a feeling for nuances in colour.

The examples on page 21 middle show how young and fresh a large room can appear using strong colours. This gave rise to the cogent title (page 25). Exuberance occasionally led to experiments, such as might have been proposed by Victor Vasarély (page 21 below). Certainly not a suggestion to last an eternity.

Then, working together with others, Peter Maly developed the first »Schöner Wohnen« colour scale, with Sahara yellow as the key colour. Daring producers developed accessory furniture, luminaires, curtains, carpets, tablecloths, paints, wallpaper, tiles and kitchen facades in these colours. If the hues matched, such objects were used in the graphic examples and the producers were named in the magazine.

At first gingerly, then later more boldly, the interior architects tackled the problem of altering established patterns for rooms wrestling with an often botched-up reality. Flowing transitions between rooms created pleasant breadth, especially in small flats. »Towers« helped subdivide rooms and left the walls intact. Using only a few colours helped avoid a gaudy spectacle.

Linke Seite

Peter Malys »Serie 1«, Tecta, 1967. Ein fröhlicher Reigen junger, kesser Ergänzungsmöbel, die das alltägliche Einerlei beleben sollten. Pop- und Op-Art hatten Farben und Bemalung angeregt, das Gespür für praktische Metaphern die dekorativen Klappbeschläge, der Wunsch nach Mobilität die Rollen unter den Kastenmöbeln. »abitare« schrieb: »... rollen, wohin du möchtest«. Kubische, richtungslose Gebilde, geeignet als Solitäre und als »Bausteine«.

On the left

Peter Maly's »Serie 1«, Tecta, 1967. A cheerful sequence of young, saucy accessory furniture, intended to liven up the monotony of daily life. Pop and Op Art had inspired colour and painting schemes; a feel for practical metaphors yielded the decorative hinges; while the desire for mobility led to the notion of putting the container furniture on casters. As the magazine »abitare« wrote »...roll whither your heart takes you«. These entities are cubic in form and without any fixed direction, suitable as free-standing units or as a »building block« for a furniture system.

nete. Ebenfalls problematisch war es für Laien, die richtige Farbmenge auszuloten. Oft ließ Begeisterung zu tief in die Farbtöpfe greifen. Die auf Seite 19 unten gezeigten Beispiele waren Versuche, den Lesern ein Gefühl für Farbnuancen zu geben. Beispiele – etwa auf Seite 21 Mitte – zeigen, wie jung und frisch starke Farben einen großen Raum erscheinen lassen können. Daraus entstand der plakative und erfolgreiche Titel (Seite 25). Gelegentlich führte Überschwang auch zu Experimenten, wie sie von Victor Vasarély angeregt sein mochten (Seite 21 unten). Gewiß kein Vorschlag für die Ewigkeit.

Je inspirierender Farbanregungen gelangen, desto mehr Leser wollten die Vorschläge für die eigenen vier Wände aufgreifen. Aber die Suche in Geschäften war oft erfolglos, denn vieles gab es gerade einmal in Hamburg, Köln oder München. Ansonsten suchte man vergeblich – nichts als genormte Langeweile.

Dann entwickelte Peter Maly mit anderen die erste »Schöner Wohnen«-Farbskala. Leitfarbe war Saharagelb in einem genau abgestimmten Farbfächer – mit Beispielen für mannigfaltige Farbklänge. Mutige Hersteller entwickelten Beimöbel, Leuchten, Vorhänge, Teppiche, Tischwäsche, Farben, Tapeten, Fliesen und Küchenfronten in diesen Farben. Stimmten die Farbtöne überein, wurden solche Gegenstände in den Bildbeispielen verwendet und Produzenten im Heft genannt.

Erst zaghaft, später mutiger gingen die Innenarchitekten daran, verfestigte Raumzuschnitte zu verändern. Sie zauberten nicht einfach Kulissenarchitektur, quälten sich vielmehr mit der oft murksigen Realität. Besonders in kleinen Wohnungen erzeugten fließende Raumverbindungen angenehme Weite. »Wohntürme« halfen Räume gliedern und ließen Wände Wände bleiben. Die Verwendung weniger Farben vermied buntes Spektakel.

Bei Umbauvorschlägen änderte Maly Raumzuschnitte nachhaltig mit Podesten als zusätzlichem Stauraum sowie mit abgehängten Decken, die gemütlichere Proportionen schufen (Seite 22 »mit Verstand umbauen«).

Weit kühner ein Vorschlag (Seite 17 auch Titel 4/68), in welchem Wandrundungen, eingezogene Deckenteile, Podeste und Lichtführung eine geradezu galaktische Raumplastik schufen.

Kunststoffe (Seite 23) schienen in den sechziger Jahren das Wohngefühl gänzlich ins Futuristische zu verwandeln. Designer wie Verner Panton, Mario Bellini, Pierre Paullin, Marco Zanuso, Eero Aarnio, Joe Colombo und viele andere erfanden Gebilde, wie man sie niemals zuvor gesehen hatte. Dieses durchgefärbte Material schien man in beliebiger Vielfalt gießen zu können. Kühle Raumplastiken, die Maly auf riesigen Treppen inszenierte (Seite 24).

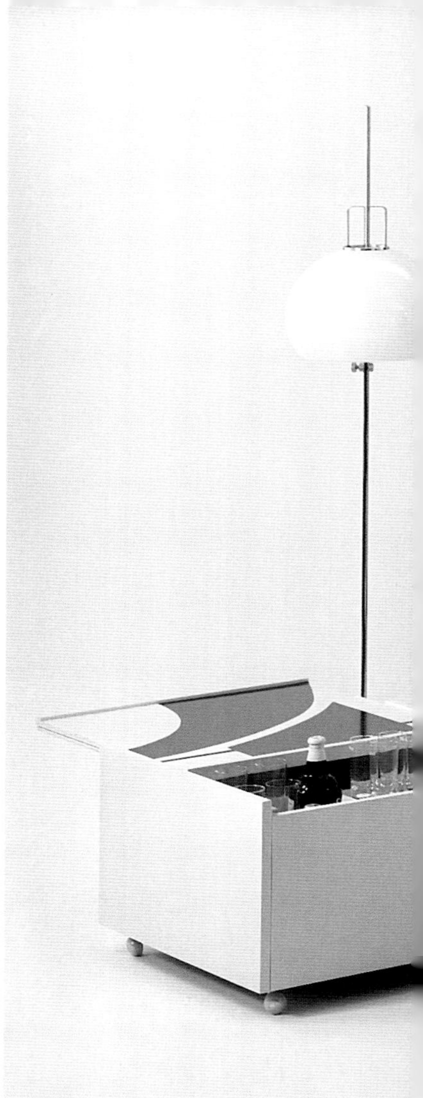

When making suggestions on how to remodel a room, Maly effected lasting changes in patterns using platforms to provide additional storage space, as well as with hung ceilings that created cozier proportions (page 22 »Intelligent Remodeling«).
Far bolder is a suggestion (page 17, issue 4/68) in which walls with rounded edges, retracted ceiling parts, platforms and lighting fixtures achieved a sense of three-dimensional space that was nearly galactic.
During the Sixties, plastics (page 23) seemed to catapult interior design completely into Futurism. Designers like Verner Panton, Mario Bellini, Pierre Paullin, Marco Zanuso, Eero Aarnio, Joe Colombo and many others invented objects the likes of which had never been seen before. It seemed that people were able to mould this material into any shape desired. Cool sculptured objects, which Maly installed on enormous staircases (page 24).
After ten years, Peter Maly left the editorial staff of »Schöner Wohnen« – although he continued to work with the magazine on a free-lance basis – to take on even greater challenges. As the »first among equals« of so many interior designers, there had been more and more for him to co-ordinate, a task which often stood in the way of his own

Gruppenbild mit Dame – »Serie 1«, Tecta, 1967. Lackiert in vier Farben: Weiß, Orange, Pink und Oliv, mit im Siebdruck aufgetragenen Dekoren.

Group Portrait with Lady –»Serie 1«, Tecta, 1967. Lacquered in four colours: white, orange, pink and olive green, with silk-screen decorations.

Nach zehn Jahren verließ Peter Maly die »Schöner Wohnen«-Redaktion – blieb jedoch ihr freier Mitarbeiter –, um sich neuen Anforderungen stellen zu können. Der primus inter pares so vieler Innenarchitekten hatte mehr und mehr zu koordinieren gehabt, eine Beanspruchung, die eigenem Entwerfen oft im Wege gestanden hatte. Auch hatten ihn zunehmend redaktionelle Wiederholungen gelangweilt, die im Jahresablauf immer wieder neue Interpretationen verlangten. Die freie Zeit außerhalb der Redaktion reichte nicht aus für eigene Designaufgaben, die ihm wichtig waren; auch standen immer mehr Hersteller vor seiner Türe.

In den sechziger Jahren war – so kann man heute im Rückblick feststellen – ein neuer Beruf geboren worden. Innenarchitekten wurden Publizisten mit vorher nicht benötigten Fertigkeiten. Peter Maly: »Die allmähliche Wandlung des Innenarchitekten zum Publizisten hatte viele Facetten: Wir hätten noch so tolle Dinge entwerfen können – wäre es uns jedoch nicht gelungen, diese zuerst der Redaktion und später auch den Lesern zu ›verkaufen‹, wäre alles vergeblich gewesen. Wir mußten rhetorisch knapp und plausibel überzeugen. Wir mußten immer bessere Zeichner werden, wir brauchten umfassende Marktkenntnisse. In Vorgesprächen mit Fotografen mußten perspektivische Zeichnungen vom Kamerastandpunkt her entwickelt werden, um den Bauaufwand ökonomisch vernünftig zu begrenzen; wir verfeinerten und schulten unser Gefühl für Licht- und Farbklänge, wir mußten lernen, uns in engem Zeitrahmen zu bewegen. In Absprachen mit Redakteuren entwickelte sich obendrein ein unerläßliches Gespür für Nuancen in Texten. In zehn Jahren lernte ich Dinge, ohne die mein weiterer Weg anders verlaufen wäre.«

capriccio für P. M.
Im Fragebogen des F.A.Z.-Magazins nennt Peter Maly als »Lieblingsblume« lakonisch Kapuzinerkresse, das Tropaeolum. Einige Kollegen, dessen bin ich sicher, haben als Antwort die duftenden Blüten der rosé rankenden Rose »New Dawn« erwartet. Jedoch, die auserkorene Lieblingsblume des Designers überzeugt: sie klettert und eignet sich zur Bekleidung hoher Wände, wird andererseits als dankbare und preiswerte Einfassungspflanze (etwa 35 Arten) bei größeren Grundstücken empfohlen. »Kapuzinerkresse« mit den schildförmigen Blättern und großen orangegelben, selten purpurnen oder blauen Blüten wird als Dekoration im Salat gegessen; die Blütenknospen und die unreifen, in Essig oder Salz eingelegten Früchte werden wie Kapern verwendet. Verschwenderisch vielseitig wie der Meister selbst.

capriccio for P. M.
In a questionnaire completed for the Friday supplement to the Frankfurter Allgemeine Zeitung, one of Germany's leading newspapers, Peter Maly laconically mentions the nasturtium, or the tropaeolum, as his »favourite flower«. I am certain that a number of colleagues were expecting him to answer with the sweetly scented blossoms of the »New Dawn« rose, with its pink tendrils. Nonetheless, the flower chosen by the designer as his favourite is convincing: it is a climbing plant, and is suitable for covering high walls; it is also recommended as a rewarding and inexpensive plant for trimming larger properties (there are around 35 varieties). »Nasturtium«, with its scutiform leaves and large, orange-yellow, or occasionally purple or blue blossoms, can be eaten as part of a salad decoration. The flower buds and the unripe fruit, marinated in vinegar or salt, are used much like capers. Extravagantly multifarious, like the master himself.

Hans Weckerle

Professor für Gestaltung an der Fachhochschule Hamburg

Professor of design at the Fachhochschule Hamburg, Germany

designs. Furthermore, he became increasingly bored by editorial redundancies, which required new interpretations throughout the course of the year. His free time outside of the editorial department did not suffice for own design tasks, and more and more producers were knocking at his door.

In retrospect, it can be stated that a new profession was born during the Sixties. Interior designers became journalists with skills that had not previously been required. In the words of Peter Maly, »There were many facets to the gradual transformation of the interior designer into a journalist. We could have designed the most wonderful things – however, if we hadn't been successful at ›selling‹ them, at first to the magazine and later to the readers as well, everything would have been in vain. We had to convince with our plausibility, and with our rhetorical conciseness. We had to become even better draftsmen, we needed a comprehensive knowledge of the market. In preliminary talks with photographers, drawings in perspective had to be developed from the point of view of the camera in order to keep construction expenses reasonable; we refined and informed our feel for shades of light and colour; we had to learn how to move in tight time frames. In addition, we developed an absolutely essential feel for nuances in texts when working with editors. Without the things I learned during those ten years, my subsequent development would have taken quite a different course.«

»Trinom«, COR, 1969. Schlichter Sessel mit Formholzschalen auf Gleitkufen. Trotz strenger Form gewährleisten Schaumgummipolsterung, gerundete Lehnen mit geringer Neigung und Abnäher bequemen Sitzkomfort. Wieder das Quadrat, passend fürs Einzelstück wie für mannigfaltige Sitzgruppierungen – eine erste Form der »Sitzlandschaft«.

»Trinom«, COR, 1969. A plain armchair with moulded wood shells on runners. Despite its austere form, it offers great seating comfort owing to its foam rubber cushioning, rounded backrest with minimal tilt, and darts. The square makes an appearance once again. This chair is appropriate as an individual piece or as part of manifold seating groups – one of the first designs of the »seating landscape«.

Peter Zec

Geometrische Ästhetik gegen visuellen Formverschleiß

Am Anfang jedes Gestaltungsprozesses steht immer wieder die Frage nach der Transformation der rein physikalischen Materialität in eine mögliche ästhetische Realität eines Gegenstands. Hierbei ist es gleichgültig, ob es sich um die Hervorhebung eines zweckfreien Kunstwerks oder eines gebrauchsfähigen Designobjekts handelt. In jedem Fall geht es darum, aus der Vielheit verfügbarer Materialien, Farben und Formen auszuwählen, umzuverteilen, neu zu ordnen und zu verwandeln, ohne daß das Ergebnis bereits von vornherein feststeht. Auch wenn es im nachhinein so aussieht, als sei es die logische Konsequenz des Gestaltungsprozesses und damit von Beginn an bekannt, kommt es während der Arbeit zumeist anders als man denkt. In diesem Akt der ästhetischen Transformation aber liegt gerade der besondere Reiz und die einzigartige Herausforderung für die Tätigkeit des Gestalters.
Drei Jahre hat es beispielsweise gedauert, bis es Peter Maly gelungen ist, die endgültige Gestaltung seines Stuhls »737« für Thonet zu entwickeln. Zum Schluß erscheint das Ergebnis so selbstverständlich, daß – wie er selbst sagt – sogar die Frage aufkom-

Geometric Aesthetics versus Visual Obsolescence Through Repetition

Every design process starts out with the question of how to transform the purely physical materiality of an object into a possible aesthetic reality. This is true regardless of whether the emphasis is on creating a work of art that serves no practical purpose, or a design object intended for use. In any case, it is a matter of selecting from the abundance of available materials, colours and forms, and then redistributing, rearranging and transforming, without knowing beforehand what the result will be. After the fact, the finished product may look as if it were the logical consequence of the design process – as if the result were obvious from the start. While the work is in progress, however, things usually do not go the way one expects. Nonetheless, it is precisely this act of aesthetic transformation which holds a special fascination for the designer and poses the unique challenge inherent in the designer's work.

It took Peter Maly three years, for example, to complete his design for the »737« Thonet chair. The final product seems so self-evident that – as he himself says – one might even be tempted to ask: »So, where is the design?« This leads to the next fundamental question: What makes design design? A question which has repeatedly led to heated debate, and which has not yet been resolved. Indeed, it would appear impossible to limit the special, defining quality of design to a specific way of looking at or interpreting things, or to any related plan for action, although there have been enough attempts to do so. At the top of the list, let us not forget Sullivan's formula, »Form

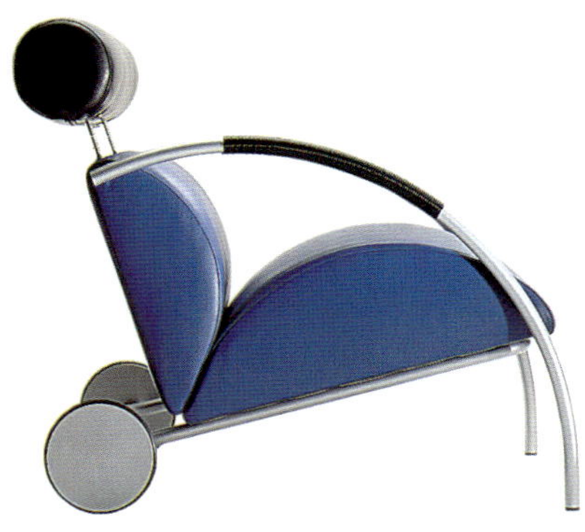

Sessel »Zyklus«, COR, 1984

»Zyklus« armchair, COR, 1984

men könnte: »Wo ist da eigentlich das Design?«. Im Grundsatz verbindet sich hiermit die darüber hinausgehende Frage: Was macht Design zum Design? Eine Frage, die in der Vergangenheit immer wieder zu heftigen Kontroversen geführt hat, ohne daß man bislang zu einem Ergebnis gekommen ist. Vielmehr scheint es unmöglich zu sein, die Besonderheit des Designs auf eine spezifische Sicht- und Deutungsweise sowie ein dazugehöriges Handlungskonzept zu beschränken, selbst wenn es genügend Versuche dieser Art gegeben hat. Allen voran ist hierbei die Sullivan-Formel des »form follows function« zu nennen, die für lange Zeit das Denken über die Bestimmung des Designs nachdrücklich beeinflußt hat.

Der Akt der ästhetischen Transformation wird auf einen Abbildungsrealismus reduziert, bei dem sich die darzustellende Form unmittelbar aus der technischen Funktion ergibt. Später wurde dieser Funktionalismus noch um die Determinante des Gebrauchs als einer grundlegenden gestalterischen Bestimmung erweitert. Dieser Auffassung gemäß gibt es keine Ästhetik an sich, die als ästhetische Realität von der physikalischen Realität eines Objekts zu unterscheiden wäre. Vielmehr liegt das Ästhetische in der Sache selbst, wo es sich durch den Gebrauch derselben zu erkennen

Linke Seite
Kommode »Serie 6«, Tecta, 1976

Tisch »M 70«, Tecta, 1977

Rechte Seite
Kommoden »Serie 7«, Tecta, 1977

On the left
»Serie 6« chest of drawers, Tecta, 1976

»M70« table, Tecta, 1977

On the right
»Serie 7« chests of drawers, Tecta, 1977

follows function«, which for a long time had a strong influence on people's thoughts about how to define design.

The act of aesthetic transformation is reduced to a representational realism in which the formal presentation of the object is a direct result of its technical function. Later, this functionalism was expanded to include the factor of utility as a fundamental component in defining design. This view states that there is no such thing as aesthetics in the sense of an aesthetic reality that can be distinguished from the physical reality of an object. Rather, the aesthetic aspect is intrinsic to the thing itself, and is revealed through use of the thing. »Design«, as Otl Aicher has put it, »dispenses with the aesthetic absolutism of art and seeks the aesthetics of use.« Without a doubt, this strips useful objects of any aesthetic autonomy or identity of their own. Optimizing their intended function is all that counts, and not expanding their intrinsic functional logic by the dimension of an aesthetic order. The beautification of the world is thus to be regarded as an improvement in the utility function of all objects existing within it. According to Otl Aicher's view, we must »see how good or how bad we are through our use of objects«.

In his designs, Peter Maly quite consciously takes issue with this pure aesthetics of utility by giving his objects an unmistakable appearance all their own. At the same time, they have a communicative aura about them. A chair such as the »737« designed for Thonet, for example, is not merely an object of everyday use in Maly's view, but

»Wenn einer dann deit, wat hei deit, dann kann hei nich mir dann, as hei deit.«

Axel Bruchhäuser

Inhaber und Geschäftsführer von Tecta in Lauenforde

Owner and General Manager of Tecta in Lauenforde, Germany

Linke Seite
Sammlerschrank »Taxus«, Reim Interline, 1978

Leseleuchte »MY«, Hersteller Anta, Vertrieb Reim Interline, 1980

Rechte Seite
Vitrinenschränke »Quadrat«, Reim Interline, 1980

On the left
»Taxus« collector's cabinet, Reim Interline, 1978

»MY« reading lamp produced by Anta. Distributed by Reim Interline, 1980.

On the right
»Quadrat« glass cabinets, Reim Interline, 1980

also a communicative object. Aside from the fact that it is quite useful and makes for very comfortable sitting, it is a piece of furniture which conveys that a great deal of importance has been assigned to a simple, nearly archetypal appearance which dispenses with superfluous decoration. Beyond its purely practical function, the chair is an expression of a specific world-view which involves the symbiosis of high quality, satisfactory comfort and a modest lifestyle.

The original formal language of the chair makes it very clear that it represents one specific aesthetic solution among many other possible solutions. This is also what gives it a special artistic quality, raising it from being merely an object of everyday use to the status of an aesthetic object, and thus imbuing it with an aesthetic value which transcends its material and utility value.

The chair therefore takes on the added function of a symbol, which in turn becomes an agent of multiple meanings. On the one hand, it is still a simple object of everyday use. On the other, it acquires the meaning of a cultural object and becomes the symbol, as it were, of a specific view of design and of life.

gibt. »Design«, so sagt Otl Aicher, »verzichtet auf den ästhetischen Absolutismus der Kunst und sucht die Ästhetik des Gebrauchs.« Zweifelsohne wird der Gebrauchsgegenstand auf diese Weise vollständig seiner ästhetischen Autonomie und Identität beraubt. Was allein zählt, ist die Optimierung seiner Zweckfunktion und nicht die Erweiterung seiner immanenten funktionalen Logik um die Dimension einer ästhetischen Ordnung. Die Verschönerung der Welt ist somit in einer Verbesserung der Gebrauchsfunktion aller in ihr existierenden Gegenstände zu sehen. Gemäß der Auffassung Otl Aichers müssen wir »im Gebrauch sehen, wie gut, wie schlecht wir sind«.

Ganz bewußt stellt sich Peter Maly mit seinen Entwürfen dieser reinen Gebrauchsästhetik entgegen, indem er den Dingen ein unverwechselbares, ihnen eigenes Aussehen verleiht, womit sie zugleich eine kommunikative Ausstrahlung erlangen. Ein Stuhl wie etwa der »737« für Thonet ist für ihn nicht allein Gebrauchsgegenstand, sondern auch Kommunikationsobjekt. Abgesehen davon, daß er gut zu gebrauchen ist und man bequem auf ihm sitzen kann, vermittelt das Möbel, daß er sehr viel Wert auf eine einfache, fast archetypische Erscheinung legt, die auf überflüssigen Dekor verzichtet. Über seine bloße Gebrauchsfunktion hinaus vermittelt der Stuhl den Ausdruck einer spezifischen Weltanschauung, die eine Symbiose von hohem Qualitätsanspruch, angemessenem Komfort und bescheidener Lebensführung zum Inhalt hat.

Peter Malys Arbeiten sind eben nicht Zeitgeist pur. Formale Tendenzen sind zögerlich und »wohl verdaut« in seine Arbeiten eingeflossen – wobei er immer seiner konsequent klaren Auffassung von Design treu bleibt.
Kompliment für diese durchgängig eindeutige Ästhetik!

Peter Maly's works are simply not pure *Zeitgeist*. Formal trends have made their way into his works only slowly and in a »well-digested« state – and he always remains true to his consistently clear convictions concerning design.
My compliments to his thoroughly unequivocal aesthetics!

Jan Wichers

Innenarchitekt in Hamburg

Interior designer in Hamburg, Germany

The tasks of the industrial designer are clearly defined, as Peter Maly sees it. For him, it is a matter of developing innovative, intelligent, high-quality products which meet the needs of target groups and are original in design. The goal is to reach as many people as possible and thus be successful in the market.

In today's world, the mere usefulness of a product is no longer sufficient to ensure that this goal will be met. In a society characterized by excess and wastefulness, the utility value of an object undergoes a transformation. It is no longer measured in terms of its usefulness and the degree to which it meets the needs it is intended to serve, but in terms of its potential for obsession. It is not a question of what shortages can be eliminated by using or purchasing a thing, but rather the benefit a thing holds in store in terms of enhanced pleasure and intensity of experience. This is becoming an increasingly important factor in selling products. Under these conditions, it is becoming more and more difficult to gain a foothold in the market. Many producers and designers today opt for a spectacular, superficial product cosmetics, which then very quickly loses its appeal and leads to accelerated visual obsolescence in the design. As a consequence, a growing number of things are being thrown away at an ever faster rate, although they could definitely still be used. It is merely that their design is no longer *en vogue*.

Linke Seite
Schlafzimmer-Ensemble »Mutaro«, Interlübke, 1979

Rechte Seite
Solitärmöbel-Programm »Art Deco«, Reim Interline, 1980

On the left
»Mutaro« bedroom ensemble, Interlübke, 1979

On the right
»Art Deco« unique furniture program, Reim Interline, 1980

Linke Seite
Bar »Duo«, Interlübke, 1986

Schreibtisch »Duo«, Interlübke, 1986

Rechte Seite
Funktionscontainer »Duo«, Interlübke, 1990

On the left
»Duo« bar, Interlübke, 1986

»Duo« desk, Interlübke, 1986

On the right
»Duo« functional container, Interlübke, 1990

In view of this less than promising trend, Peter Maly consciously bases his design on the principle of slowing down the process of formal obsolescence. He is mindful in particular to use basic geometric forms as the point of departure and the basis for his designs. The Hamburg designer is thus part of that long tradition of design thinking which goes back to Plato. In his »Timaeus«, Plato expounds on the worldly and cosmic order by referring to geometric bodies. The geometric aesthetic which goes back to his world-view essentially follows the principle of logos or ratio, guided by proportion and order. The laws of proportion (nomoi), such as the Golden Section, define the shape of the design. The goal of this aesthetic is to lend proportion and order to the chaotic and coincidental elements of the visible world using mathematical equations and geometric figures.
Luca Pacioli takes up this principle again in the 15th century in order to stress the importance of cosmic harmonies as a regulative principle in both music and architecture. »It is impossible«, he says, »for anything to exist in nature if it is not proportioned according to its nature.« The human being is ultimately also integrated into this system, inscribed in a circle and a square – without these basic forms, it is not possible to design anything at all.
Because the geometric aesthetic is elevated to a cosmic universality, it is given a valency and purpose which enables it to transcend the realm of temporal transience, and thus stand for the creative principle of nature. This creative principle is applied to all artistic design processes by virtue of the geometric aesthetic.
Max Bense also speaks of this in his »Aesthetica« when he speaks in terms of aesthetic reality as distinct from the physical reality of a design object. »Aesthetic reality appears here as a second aspect of cosmological reality, in addition to physical reality. Aesthetic

Lieber Peter!
– Der Bleistift ist, wie ich seit Beginn unserer Zusammenarbeit, Anfang der siebziger Jahre, beobachten konnte, Dein primäres (sprich: wichtigstes) Handwerkszeug.
– Die Hasselblad und das 50 mm Objektiv waren und sind bis heute unser gemeinsames Handwerkszeug.
– »Schöner Wohnen«, eine deutsche Wohnzeitschrift, leistete sich seinerzeit aufwendige Fotoproduktionen mit erkennbarer Handschrift der Architekten und Fotografen.
– Für mich zählte damals nur italienisches Möbeldesign. Heute gibt es auch Möbeldesign von Peter Maly.
– Ich selbst besitze von Dir einen Bistro-Tisch und den ersten »Zyklus«, der produziert wurde.
– Wir arbeiten immer noch zusammen, was mich ungemein freut und worauf ich sehr stolz bin.

Dear Peter,
– As I have been able to observe since we began working together at the beginning of the Seventies, the pencil is your primary (read: most important) tool.
– We have always had the Hasselblad camera and the 50mm objective lens in common.
– »Schöner Wohnen«, a German interior decor magazine, used to indulge in photographic productions which for were extravagant for the times, and in which the artistic styles of the architects and photographers were recognisable.
– In those days, I only had eyes for Italian furniture design. Today there is also Peter Maly furniture design.
– I myself own your Bistro table and the first Zyklus chair ever produced.
– We still work together, much to my delight – something of which I am very proud.

Rudolf Schmutz jr.

Fotograf in Hamburg

Photographer in Hamburg, Germany

Aufgrund seiner originären Formensprache verdeutlicht das Sitzmöbel, daß es eine spezifische ästhetische Lösung von vielen anderen möglichen Lösungen darstellt. Hierin liegt zugleich die besondere gestalterische Qualität, die es vom bloßen Gebrauchsgegenstand in den Stand eines ästhetischen Objekts erhebt. So wird ihm ein über den Material- und Gebrauchswert hinausgehender ästhetischer Wert verliehen. Zusätzlich übernimmt der Stuhl also die Funktion eines Zeichens, welches zum Träger mehrerer Bedeutungen wird. Zum einen stellt er nach wie vor einen schlichten Gebrauchsgegenstand dar. Zum anderen nimmt er die Bedeutung eines Kulturobjekts an und wird gleichsam zum Symbol für eine spezifische Gestaltungs- und Lebensauffassung.

Das Aufgabenfeld des Industrie-Designers ist für Peter Maly klar umrissen. Für ihn geht es um die Entwicklung innovativer, intelligenter, zielgruppen-adäquater und eigenständig gestalteter Produkte von hoher Qualität. Ziel ist, möglichst viele Menschen zu erreichen und so im Markt erfolgreich zu sein.

Die bloße Brauchbarkeit eines Produkts reicht heute nicht mehr aus, um dieses Ziel sicher zu erlangen. In einer durch Überfluß und Verschwendung gekennzeichneten Gesellschaft erfährt der Gebrauchswert eines Gegenstands einen Bedeutungswandel. Er wird nicht länger an der Qualität der bedarfsgerechten Handhabung und Nützlich-

Oben
Couch »Prao«, Ligne Roset, 1985

Unten
Schranksystem »Onde«, Ligne Roset, 1991

Above
»Prao« couch, Ligne Roset, 1985

Below
»Onde« cupboard system, Ligne Roset, 1991

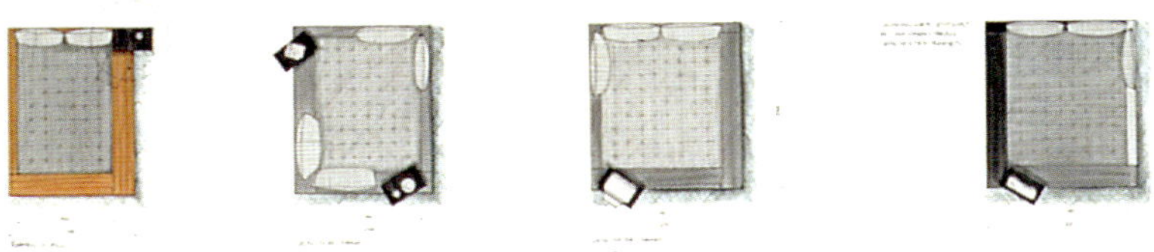

Von oben nach unten

Recamiere »Antares«, Ligne Roset, 1988

Bett »Peter Maly«, Ligne Roset, 1983

Entwurf zum »Peter Maly«-Bett, Ligne Roset, 1983

From top to bottom

»Antares« day bed, Ligne Roset, 1988

»Peter Maly« bed, Ligne Roset, 1983

Draft of the »Peter Maly« bed, Ligne Roset, 1983

Ein Zitat von Mies van der Rohe charakterisiert Peter Maly am treffendsten: »Less is more«. Er ist Purist bis ins letzte Detail und nicht bereit, auf halbem Weg Kompromisse einzugehen. Seine klaren, geometrisch begründeten Formen sind immer perfekt aufeinander abgestimmt. Sie sind von zeitloser Eleganz, niemals beliebig. Das und sein untrüglicher Sinn für harmonische, reduzierte Farb- und Formgebung machen ihn zu einem unserer wichtigsten Kreativen.
Viele charakteristische Modelle unserer Kollektion stammen von ihm, auch unser Kölner Messestand und das Interieur führender Ligne Roset-Exklusivgeschäfte.

A famous statement by Mies van der Rohe most aptly characterizes Peter Maly: »Less is more«. He is a purist down to the last detail, and he is not prepared to make compromises. His clear, geometrically based forms are always perfectly coordinated with one another. They possess a timeless elegance, and are never arbitrary. This, combined with his unerring feel for harmonious, reduced colours and forms make him one of the most important figures in the creative world. Many of the models which are characteristic of our collection were designed by Maly, including our exhibits at the Cologne trade fair, and the interiors of Ligne Roset shops.

Pierre Roset

Geschäftsführer von Ligne Roset in Serrieres-de-Briord, Frankreich

General Manager of Ligne Roset in Serrieres-de-Briord, France

keit, sondern an seinem Obsessionspotential gemessen. Nicht die Frage, welche Mangelsituation sich durch den Gebrauch bzw. Kauf einer Sache beseitigen läßt, sondern die Frage nach dem Lustgewinn und der Erlebnisintensität, den bzw. die eine Sache verspricht, gewinnt für den Absatz von Produkten an Bedeutung. Unter diesen Bedingungen wird es immer schwerer, sich im Markt zu behaupten. Häufig wird heute auf eine spektakuläre, oberflächliche Produktkosmetik gesetzt, die dann sehr schnell an Attraktivität verliert und zu beschleunigtem visuellem Formverschleiß führt. Dies hat zur Folge, daß immer mehr Dinge immer schneller entsorgt werden, obwohl sie durchaus noch gut zu gebrauchen wären, allein ihre Form entspricht nicht mehr der aktuellen Mode.

Angesichts dieser wenig verheißungsvollen Entwicklung setzt Peter Maly mit seinem Design gezielt auf das Prinzip der Verlangsamung des Formverschleißes. Hierbei besinnt er sich vor allem auf geometrische Elementarformen als Ausgangspunkt und Basis seiner Entwürfe. Damit reiht sich der Hamburger Designer in jene lange Tradition des gestalterischen Denkens ein, die ihren Ausgangspunkt bei Platon hat. Der Philosoph erläutert in seinem »Timaios« die irdische und kosmische Ordnung mit Hilfe geometrischer Körper. Die auf Platons Weltanschauung zurückgehende geo-

Linke Seite
Entwurf zu »Zyklus«, COR, 1984

Detail der limitierten Auflage des Sessels »Zyklus«, COR, 1984/94

Rechte Seite
Oben
Sessel »Zyklus«, COR, 1984

Mitte rechts
Entwurf zu »Cirrus«, COR, 1992

Mitte links
Sessel und Sofa »Zyklus«, COR, 1984

Unten
Sitzliege »Cirrus«, COR, 1992

On the left
Draft of »Zyklus«, COR, 1984

Detail of the limited edition »Zyklus« armchair, COR, 1984/94

On the right
Above
»Zyklus« armchair, COR, 1984

Middle right
Draft of »Cirrus«, COR, 1992

Middle left
»Zyklus« armchair and sofa, COR, 1984

Below
»Cirrus« lounge chair, COR, 1992

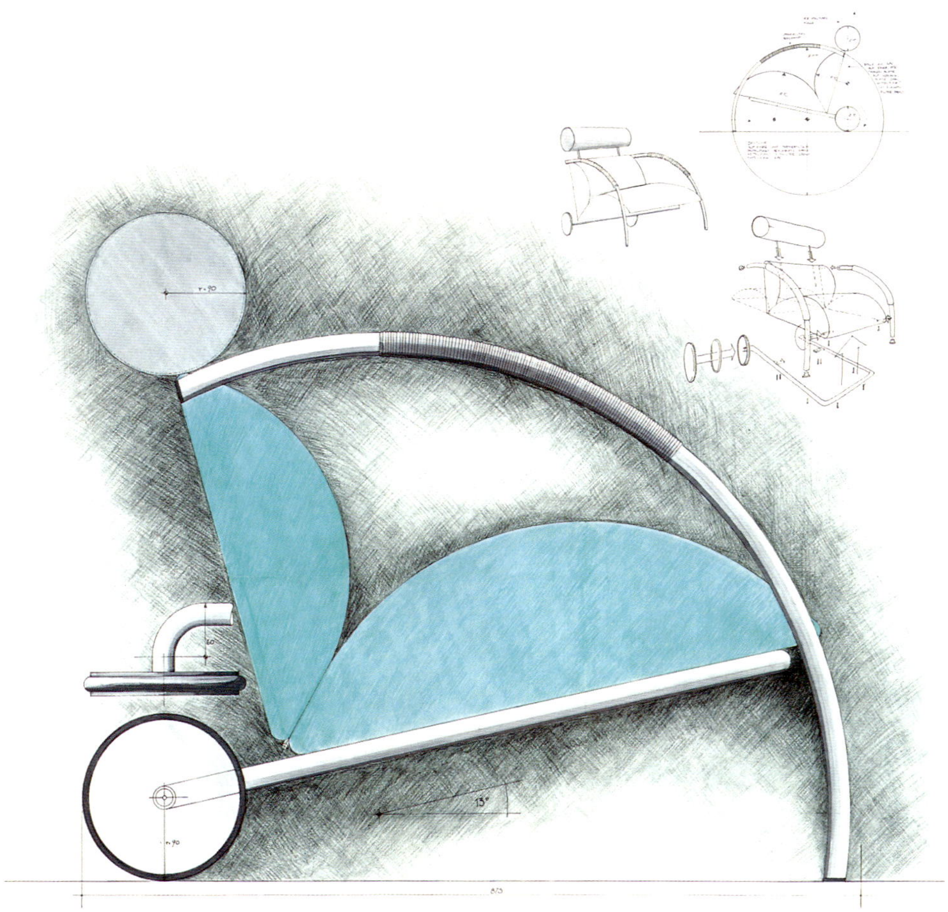

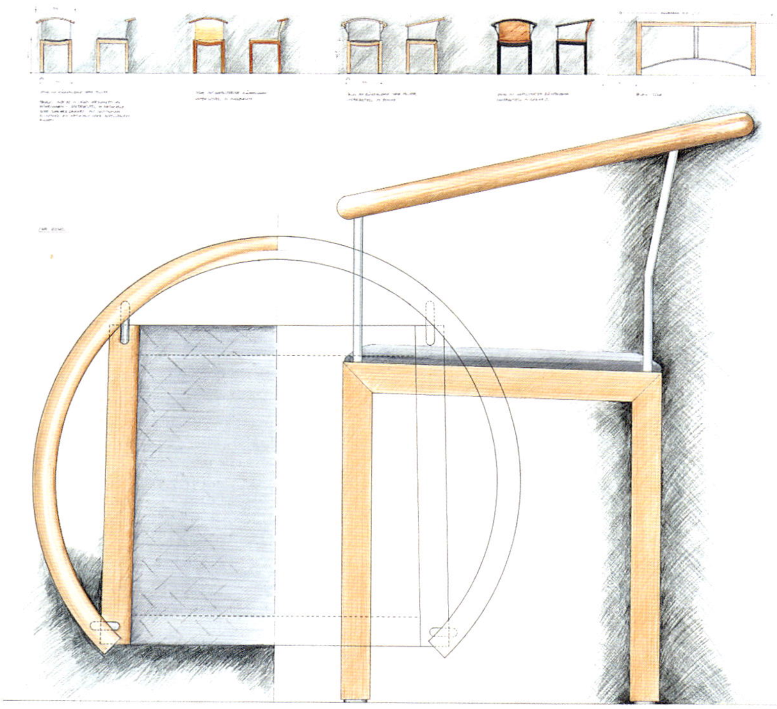

metrische Ästhetik folgt im wesentlichen dem Prinzip des nach Maß und Ordnung ausgerichteten Logos bzw. der Ratio. Dabei sind die Gesetze der Maßverhältnisse (nomoi), wie zum Beispiel das des Goldenen Schnitts, für die Proportionen der Gestaltung bestimmend. Ziel dieser Ästhetik ist es, das Chaotische und Zufällige der sichtbaren Welt durch mathematische Formeln und geometrische Figuren zu proportionieren und zu ordnen.
Im 15. Jahrhundert greift Luca Pacioli dieses Prinzip wieder auf, um damit die Bedeutung der kosmischen Harmonien als ein regulatives Prinzip sowohl der Musik als auch der Architektur hervorzuheben. »Es ist unmöglich«, so sagt er, »daß etwas in der Natur bestehen könnte, wenn es nicht seiner Natur gemäß proportioniert wäre.« In dieses System wird schließlich auch der Mensch integriert. Er wird in einen Kreis und ein Quadrat eingeschrieben, ohne deren Grundformen es nicht möglich ist, überhaupt etwas zu entwerfen.

»Man braucht doch keinen Stuhl zu erfinden. Den gibt es. Was man tun kann, beschränkt sich darauf, ihn bequem, einladend, solid, von edlen Verhältnissen zu bilden und sachgerecht zu schmücken. Innerhalb dieser Grenzen sind die Möglichkeiten Legion.« Soweit Alfred Lichtwarks »Brief an die Commission« von 1901.
Designer sehen sich heutzutage zwangsläufig in der Pflicht, die Welt des Scheins zu befriedigen, neue Produkte zu kreieren, die die Welt edler, dynamischer, jünger erscheinen lassen, und damit Bedürfnisse zu wecken, von denen heute kaum jemand ahnt, daß sie morgen gefragt sind. Der Designer muß diese Arbeit ernst und wichtig nehmen; nur sie garantiert ihm Aufträge und Erfolg.
Peter Maly indes kann und will mehr. Auch er unterliegt dem Gesetz sich ständig abwechselnder Novitäten, doch in seinem Innersten bleibt er Perfektionist. Peter Maly, von Hause aus Tischler, liebt das Material, hat Respekt vor dem Holz. Anzunehmen ist, daß Shaker-Möbel ihn begeistern und seinen Wunsch nach zeitloser, anonymer Schönheit erfüllen.
Malys Entwurf eines Stuhls für Thonet aus dem Jahr 1995 ist diesem Ideal verpflichtet: schmucklos einfach, mit vier geraden Beinen, Sitz und Lehne mit Gurten bespannt. Dieser Stuhl strahlt Leben und Wärme aus, steht in der Tradition skandinavischer Stuhlentwürfe.
Das Ergebnis ist bewußt keine Sitzskulptur. Was Maly will, ist eine formale Behandlung, die unauffällig bleibt, aber im Detail stimmt. Im Zusammenspiel der Proportionen liegt der Gewinn. Dieser Stuhl ist leicht und freundlich, einladend und nach kurzer Zeit vertraut.
Unter den obwaltenden wirtschaftlichen Umständen wird das Gute immer wieder umgestoßen und durch Neues verdrängt. Vielleicht sind Designer wie Peter Maly dazu da, dafür Sorge zu tragen, daß immer wieder Gutes nachkommt.

»No one has to invent the chair. That has already been done. What remains is limited to making the chair comfortable, inviting, solid, exquisitely proportioned, and decorating it properly. Within these limits, the possibilities are manifold.« Alfred Lichtwark: »Brief an die Commission«, 1901.
Nowadays, designers feel compelled to satisfy the world of semblance, to create new products that make the world seem more noble, more dynamic, and younger, and to awaken new desires on the part of consumers. Few can predict today what will be in demand tomorrow as a result of this process. The designer must consider these tasks important and take them seriously, as this alone is what brings in work and makes for success.
Whereas Peter Maly can and wants to do more. He, too, is subject to the law of the latest fashion, but he remains a perfectionist at heart. Peter Maly was originally trained as a cabinetmaker; he loves the material and respects the wood he works with. Presumably, he is inspired by Shaker fur-

conditions, at least in principle, are cosmological in nature in the same sense as physical conditions are. Art or design can thus be defined as intelligent beings intervening in the physical conditions of cosmological reality in order to achieve aesthetic conditions.« Bense refers to the way in which this happens as the process of aesthetic transformation.

Peter Maly executes his creative work in the awareness of this aesthetic theory, which he puts into practice in his designs and products. An especially vivid example is the »Zyklus« easy chair, the product of a design created for the COR firm in 1984, which has meanwhile become a classic. It is far more than a use-oriented piece of furniture to sit on. The product derives its added, aesthetic value from the consistent application of the geometric theory of proportion, with the circle providing the basic form in this

niture, which satisfies his desire for timeless, anonymous beauty.

The chair which Maly designed for Thonet in 1995 is indebted to this ideal: unadorned and simple, with four straight legs, the seat and backrest covered with woven fabric. In the tradition of Scandinavian chair design, this piece of furniture exudes life and warmth.

The result is consciously not a seating sculpture. What Maly intends is a formal treatment that remains inconspicuous but which is perfect down to the last detail. What the chair lacks in adornment, it makes up for with its interplay of proportions. It is lightweight, friendly and appealing, and makes one feel at ease very quickly.

Under the prevailing economic circumstances, quality is constantly being overturned and pushed back by newness. Perhaps designers like Peter Maly are here to ensure that quality will always make a comeback.

Dr. Rüdiger Joppien

Kustos am Museum
für Kunst und Gewerbe
in Hamburg

Curator at the Museum
for Arts and Crafts
in Hamburg, Germany

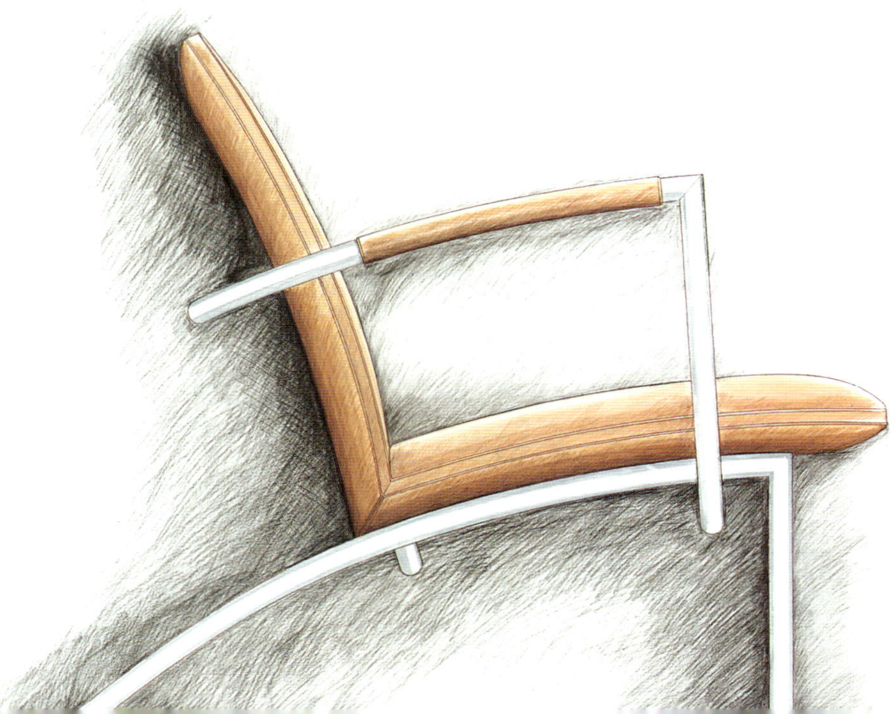

Linke Seite
Stuhl- und Tischfamilie »Taurus«, COR, 1990

Entwurf zu »Taurus«, COR, 1990

Rechte Seite
Einzelsessel »Pando«, COR, 1992

Entwurf zu »Pando«, COR, 1992

On the left
»Taurus« table and chair group, COR, 1990

Draft of »Taurus«, COR, 1990

On the right
»Pando« armchair, COR, 1992

Draft of »Pando«, COR, 1992

Linke Seite, von oben nach unten
Couchtisch »Tarbo«, COR, 1990

Einzelsessel »Logo«, COR, 1988

Entwurf zur Tischserie »Zelda«, COR, 1990

Rechte Seite, von oben nach unten
Sitzmöbel-Elemente »Tarbo«, COR, 1990

Polsterbett »Tarbo«, COR, 1991

Sessel »Zelda«, COR, 1988

On the left, from top to bottom
»Tarbo« coffee table, COR, 1990

»Logo« armchair, COR, 1988

Draft of the »Zelda« table series, COR, 1990

On the right, from top to bottom
»Tarbo« seating elements, COR 1990

»Tarbo« upholstered bed, COR, 1991

»Zelda« armchair, COR, 1988

Peter Maly zähle ich zu den wenigen Möbelgestaltern mit internationaler Geltung. Mit seinem klaren und nachvollziehbaren Gestaltungskonzept gibt er den Gegenständen eine langlebige Formbeständigkeit. Unterstützt durch die Material- und Verarbeitungsqualität der jeweiligen Hersteller, leistet er einen erheblichen Beitrag zu unserer Wohnkultur und nicht zuletzt zum Thema Ökologie.

I count Peter Maly among the few furniture designers of international importance. With his clear and understandable design concepts, he gives objects a form that is meant to last. Supported by high-quality material and the processing methods of the respective producer, he makes a major contribution to the way we live, and not least of all to ecology.

Jan Armgardt

Designer in Ludwigshafen

Designer in Ludwigshafen, Germany

Indem die geometrische Ästhetik zu einer kosmischen Universalität erhoben wird, erhält sie eine Wertigkeit und Bestimmung, die sie der zeitlichen Vergänglichkeit enthebt. Sie steht damit stellvertretend für das schöpferische Prinzip der Natur, das durch sie nun auch auf alle künstlerisch gestalterischen Schöpfungsprozesse übertragen wird. Hierauf bezieht sich nicht zuletzt auch Max Bense in seiner »Aesthetica«, wenn er von der ästhetischen Realität im Unterschied zur physikalischen Realität eines Designobjekts spricht. »Ästhetische Realität erscheint hier als ein zweiter Aspekt der kosmologischen Wirklichkeit neben ihrer physikalischen Realität. Ästhetische Zustände sind, wenigstens im Prinzip, im gleichen Sinne kosmologischer Natur wie physikalische Zustände. Kunst beziehungsweise Design kann dementsprechend als Eingriff intelligenter Wesen in die physikalischen Zustände kosmologischer Wirklichkeit definiert werden, um ästhetische Zustände zu gewinnen.« Die Weise, in der dies geschieht, bezeichnet Bense als den Prozeß der ästhetischen Transformation.

Das gestalterische Schaffen Peter Malys vollzieht sich im Bewußtsein dieser ästhetischen Theorie, die in seinen Entwürfen und Produkten in die Praxis übertragen wird. Ein besonders anschauliches Beispiel ist der inzwischen zum Klassiker avancierte Sessel

Linke Seite, von oben nach unten
Relax-Sessel »Sera«, COR, 1994

Sessel »Vega/Veganova«, COR, 1992/94

Couch »Sera«, COR, 1994

Rechte Seite
Einzelcouch »Sharky«, Brunati, 1992

On the left, from top to bottom
»Sera« easy-chair, COR, 1994

»Vega/Veganova« armchair, COR, 1992/94

»Sera« couch, COR, 1994

On the right
»Sharky« couch, Brunati, 1992

case. The various elements and details are not developed according to a predetermined linear process, but instead emerge in the course of what is much more an experimental approach, a search, feeling out and trying out. The circle, however, is never abandoned as the prevailing formal principle – as the pivotal, central aspect.
This is also the basis for the long-lived, timeless design which derives its effect from the interplay of the familiar circle and the unforeseeable aesthetic transformation of this shape. The special quality of the object's aesthetic message consists in shifting the statistical nature of the series of symbols based on the circle toward a higher level of improbability, a higher plane of innovation, surprise, uncertainty, the essence of which Max Bense refers to as originality.
To the extent that Peter Maly has been able to alienate the geometric form of the circle in such a way that it becomes a surprising semantic medium of the design process, he has also succeeded in creating a completely autonomous, new aesthetic reality of enduring value. At the same time, he has come up with a plausible answer to the current problem of formal obsolescence in our affluent society. As he himself says, he has an ideal formal vocabulary as an antidote to optical obsolescence. And this is also what constitutes the special nature and quality of his designs.

Ich mag Unterschiede. Nur darin besteht Design. Peter Maly ist wahrhaft professionell. Wir arbeiten für das gleiche Unternehmen, doch auf sehr unterschiedliche Weise. Nur so ist kultureller und wirtschaftlicher Reichtum möglich.

I like contrasts. After all, design exists only as a result of contrasts. Peter Maly is a true professional. We work for the same company, but in very different ways. Only in this way can we go about contributing to both cultural and economic wealth.

Pascal Mourgue

Designer in Paris, Frankreich

Designer in Paris, France

Linke Seite, von oben nach unten
Teppich Peter Maly-Edition »Medio II«, JAB Anstoetz, 1993

Teppich Peter Maly-Edition »Centercourt«, JAB Anstoetz, 1993

Interieur mit Teppich »Crossing«, JAB Anstoetz, 1993

Rechte Seite, von oben nach unten
Entwurf zum Teppich »Crossing«, JAB Anstoetz, 1993

Teppich Peter Maly-Edition »Gondwana«, JAB Anstoetz, 1993

Interieur mit Teppich »Minipoint«, JAB Anstoetz, 1993

On the left, from top to bottom
»Medio II« rug for the Peter Maly Edition series, JAB Anstoetz, 1993

»Centercourt« rug for the Peter Maly Edition series, JAB Anstoetz, 1993

Interior with »Crossing« rug, JAB Anstoetz, 1993

On the right, from top to bottom
Draft of »Crossing«, JAB Anstoetz, 1993

»Gondwana« rug for the Peter Maly Edition series, JAB Anstoetz, 1993

Interior with »Minipoint«, JAB Anstoetz, 1993

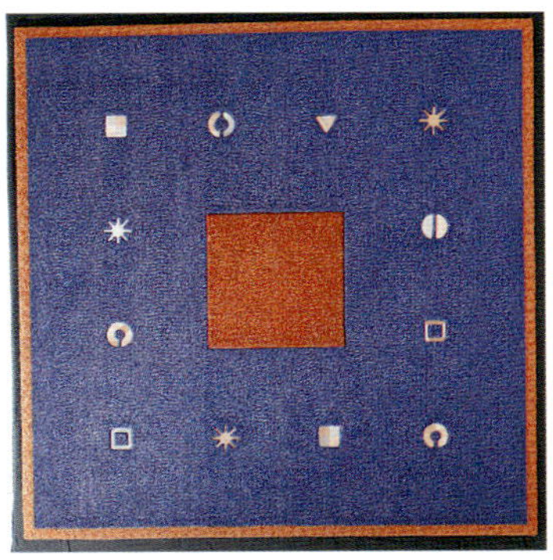

»Zyklus«, der 1984 aus einem Entwurf für die Firma COR entstand. Es handelt sich um weitaus mehr als ein gebrauchsorientiertes Sitzmöbel. Seinen ästhetischen Mehrwert bezieht das Produkt aus der konsequenten Anwendung der geometrischen Proportionslehre, der in diesem Fall der Kreis als Basisform zugrunde liegt. Dabei ist es keineswegs so, daß die Entwicklung der unterschiedlichen Elemente und Details als ein vorbestimmter linearer Prozeß erfolgt. Vielmehr geht sie aus einem experimentellen Vorgehen, einem Suchen, Herantasten, Ausprobieren hervor. Der Kreis allerdings als das formal beherrschende Prinzip – sozusagen als Dreh- und Angelpunkt – wird niemals aufgegeben.

Hierauf gründet sich auch die langlebige, zeitlose Gestalt, die ihre Wirkung aus dem Zusammenspiel der gewohnten Form des Kreises und der unvorhersehbaren ästhetischen Transformation desselben bezieht. Die Besonderheit der ästhetischen Botschaft des Objekts besteht in der Verschiebung des statischen Charakters der Zeichenfolge des Kreises in Richtung einer höheren Unwahrscheinlichkeit, höherer Innovation, Überraschung, Unbestimmtheit, deren Inbegriff von Max Bense als Originalität bezeichnet wird.

In dem Maße, wie es Peter Maly gelungen ist, die geometrische Form des Kreises zu einem überraschenden Bedeutungsträger der Gestaltung zu verfremden, ist es ihm auch gelungen, eine völlig eigenständige, neue ästhetische Realität von bleibendem Wert zu schaffen. Zugleich hat er damit eine plausible Antwort auf den aktuellen Formverschleiß in unserer Überflußgesellschaft gegeben. Wie er selbst sagt, hat er ein ideales Formenvokabular als Rezept gegen optischen Verschleiß gewonnen. Hierin liegen zugleich die Besonderheit und die Qualität seiner Entwürfe begründet.

Mit Freude arbeiten wir seit über zwanzig Jahren erfolgreich mit Peter Maly zusammen. Sein Einfühlungsvermögen und die Menschlichkeit seiner Entwürfe werden von uns und den Käufern hoch geschätzt. Ausgewogene und feine Proportionen kennzeichnen seine Modelle.
Das Programm »Quadrat« aus dem Jahre 1980 wird bereits als Klassiker gehandelt. »Basic« wurde vom Design Zentrum Nordrhein Westfalen mit dem »Roten Punkt« ausgezeichnet. Vielleicht der nächste Klassiker?

Much to our delight, we have enjoyed working fruitfully with Peter Maly for over 20 years. His intuitive understanding and the human quality of his designs are highly appreciated by our company, as well as by consumers. His models are characterized by fine, balanced proportions.
The »Quadrat« series from 1980 is already considered a classic. »Basic« was awarded the »Red Dot« by the Design Center in North Rhine-Westphalia. The next classic, perhaps?

Christine Kröncke

Geschäftsführerin von Interline Nova in München

General Manager of Interline Nova in Munich, Germany

Linke Seite

Vitrinenschränke »Quadrat Nova«, Reim Interline, 1993

Rechte Seite

Vitrinenschrank »Quadrat Nova«, Reim Interline, 1993

Sideboard-Kombination »Quadrat Nova«, Reim Interline, 1993

Vitrinentisch »Quadrat Nova«, Reim Interline, 1993

On the left

»Quadrat Nova« glass cabinet, Reim Interline, 1993

On the right

»Quadrat Nova« glass cabinet, Reim Interline, 1993

»Quadrat Nova« sideboard combination, Reim Interline, 1993

»Quadrat Nova« glass table, Reim Interline, 1993

»Zeitcollagen«/»Time collages«

Eine Auswahl von Zeichnungen, Skizzen, Arbeitsmodellen, Materialien, Fotos, Dokumentationen und Möbeln im zeitlichen Kontext

A selection of drawings, sketches, working models, materials, photographs, documentations and furniture in a chronological context.

Konzept und Arrangement/Concept and arrangement: Peter Maly

ABITARE
and these you can put where you like
OBJEKT-KUNST
Großformat
ANDY WARHOL CARS
SEKRETÄR UNI
SEKRETÄR DEKOR
BAR UNI
BAR DEKOR

TECTA
Baroni, D'Auria
Josef Hoffmann und die Wiener Werkstätte
reiminterline
ART DECO
SINGULARMÖBEL
COLLEKTION

›taxus‹ ›basis‹
interline
›mutaro‹
Interlübke

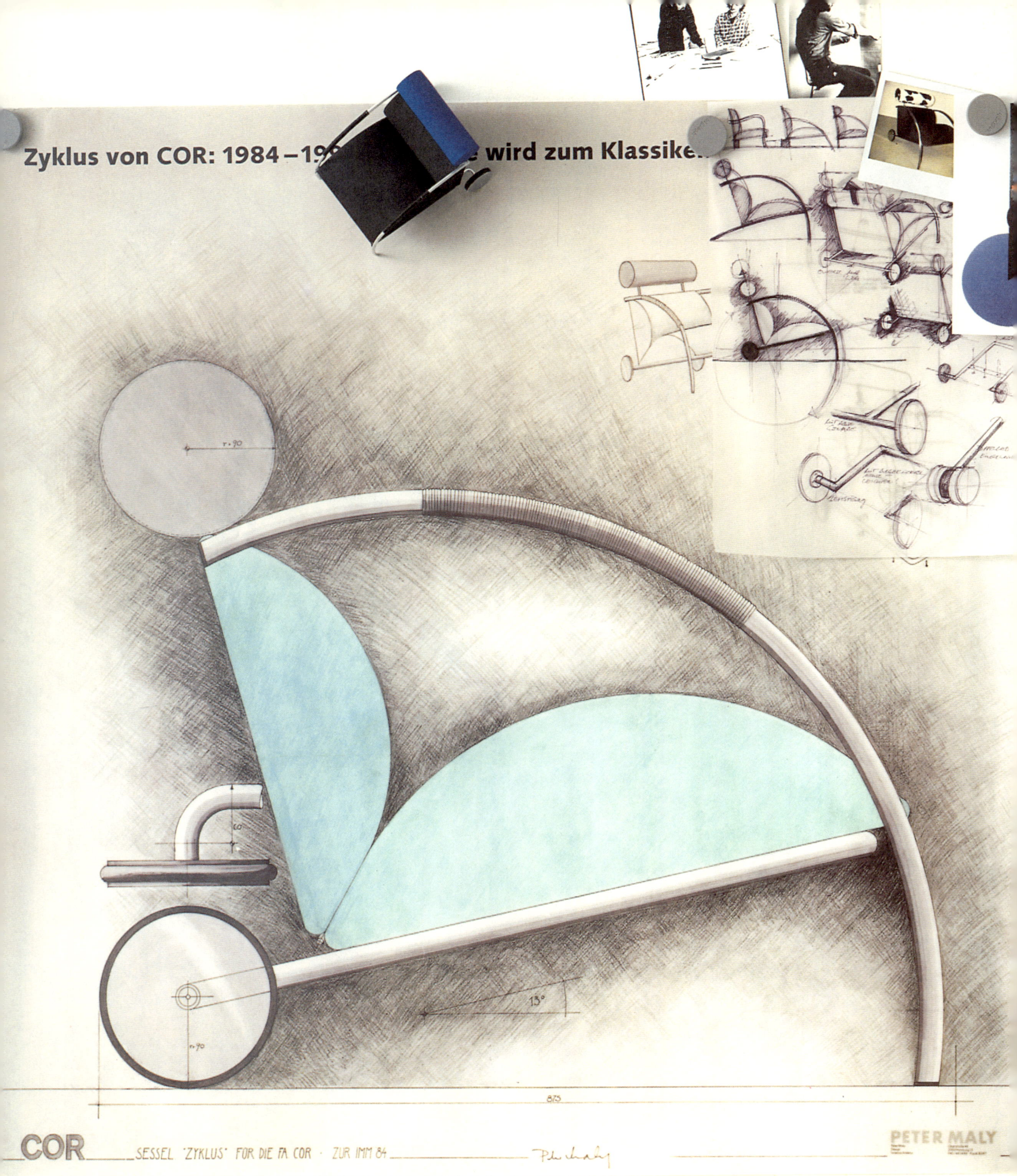

Zyklus von COR: 1984 – 19
wird zum Klassike
r. 90
60
r. 90
13°
875
COR
SESSEL "ZYKLUS" FÜR DIE FA COR · ZUR IMM 84
PETER MALY

Black and white – the perfect partnership.
interlübke
Messestand 1988

behr INTERNATIONAL
DIE IDEE WIRD ZUM PROGRAMM!
1977–1994
Möbeldesign
von Peter Maly
y – ein stilprägender Möbeldesigner Europas.
wahl aus seinem Schaffen präsentieren wir in einer
au vom 16. 4. – 28. 5. 1994 im Roland Forum, 6. Etage.
während der Geschäftszeiten.
METRIX

Die JAB TEPP
faktur mit ih
Erfahrung in de
dividuellen Fertig
Künstlerteppichen
Objekte und den
bereich präsentiert die PETER MALY
EDITION.
Diese handgearbeiteten „Kunstwerke am Boden“ dokumentieren nicht nur aktuelles Textildesign, sie bürgen auch für ein hohes Maß an handwerklichem Können. Reine Schurwolle und Seide mit in den Teppichfond eingelassenen Metallen (Kupfer, Messing, Stahl) und Leder sind die Materialien.
Teppich wird in unserer Manu-
eln von Hand in hervor-
ualität gearbeitet. Für das
Interieur können Sie des-
öße und, wenn erforderlich,
rben selbst bestimmen.
Formen,
spielen kann. Ich lö
mathematischen Sta
sie neu zusammen zu
Collagen.
Die so entstandene
keine entrückten Einz
immer als Teil des Ra
PET
Genthiner Str. 41, 10785
Im Hübner-Haus, 4.–6. Et
Telefon 030/2 54 05-20
Telefax 030/2 61 91 98
SENZA PAROLE
WITHOUT WORDS
ciuto prevalentemente come desi-
e tavoli, il tedesco Peter Maly ha
anche una serie di accessori per la
camente piacevoli prodotti dalla
il suo nome, è composta da sedi-
caffettiera, teiera, portacandele,
oi, in acciaio e legno di faggio o in
o di pero.
Good appetite! German Peter Maly is mainly known as a chair, sofa and table designer. He recently designed a range of useful and nice table accessories manufactured by Wilkens.
Cycla (the range's name) consists of sixteen steel and beechwood or Sterling silver and pear wood pieces, such as a coffee pot, a tea pot, a candlestick, a fruit bowl and trays.

reiminterline
basic
PETER MALY DESIGN
PETERMALY
ligne roset
CENA - STUHLPROGRAMM FÜR TONON

AAK Jourfixe 9/94
Di 29.November'94 19h
bei Peter Maly Design
20144, Oberstraße 46
Dokumentation einer
industr.Möbelentwicklung
Thema:"Thonet und Maly
das Thonetprogramm 737"
Es referieren Philipp Thonet und Peter Maly
· TICINO · MÖBELPROGRAMM FÜR EIN ESSZIMMER · STUHL · TISCH · BÜFFETWAGEN ·

The Sixties Die sechziger Jahre

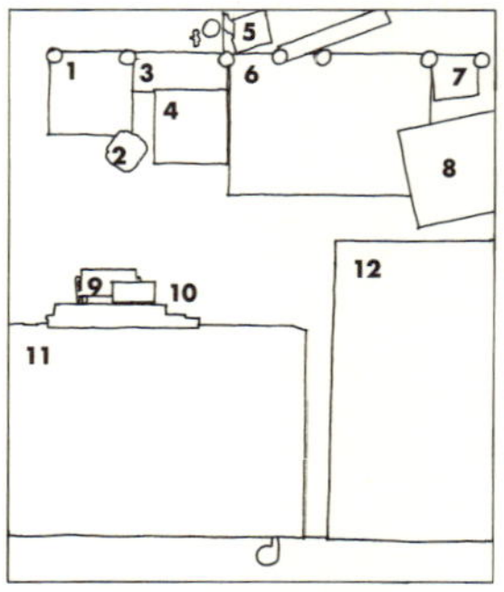

1+4 Fotos von »Trinom«, meinem ersten Entwurf für COR, 1969 2 Sesselmodell meiner Kinder 3 Entwurfsskizze – erster Messestand für COR 5 Mit zwei Kollegen von »Schöner Wohnen«: Rüdiger Zeitz und Lutz Guderian auf großem Mexico-Trip 6 Erste Veröffentlichung der »Serie 1« (Tecta) in der italienischen Zeitschrift »Abitare« (60er) 7 Quadratische Kommode der »Serie 1«, 1967 8 »Gruppenbild mit Dame« der »Serie 1«, 1967 9 Puppensessel – Ursprung des späteren »Zyklus«-Entwurfs 10 Modell 1:10 »Serie 1« 11 Original-Kommode »Serie 1«, mein erstes Serienmöbel, 1967 12 Farbentwürfe 1:10 der »Serie 1«, Dekore von M. Dupont, 1966

1 + 4 Photos of »Trinom«, my first design for COR, 1969 2 My children's chair models 3 Design sketches – COR's first trade fair exhibit 5 Trip to Mexico with two »Schöner Wohnen« colleagues: Rüdiger Zeitz and Lutz Guderian 6 First publication of »Serie 1« (Tecta) in the Italian magazine »Abitare« (the Sixties) 7 Square-shaped chest of drawers from »Serie 1«, 1967 8 »Group Portrait with Lady« from »Serie 1«, 1967 9 Doll's chair – basis for the later »Zyklus« design. 10 »Serie 1« model on a scale of 1:10 11 Original chest of drawers from »Serie 1«, my first furniture series 12 Colour designs of »Serie 1« on a scale of 1:10, decorations by M. Dupont, 1966

The Seventies Die siebziger Jahre

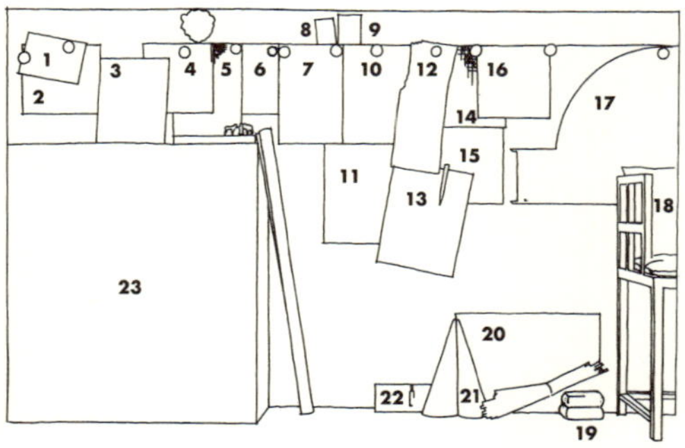

1 Foto Tisch »M 70«, 1976 (Tecta) 2 Katalog-Titel »Tecta« 3 Foto: Gisela Maly, Stylistin des »Tecta«-Standes, 1975 4 Foto aus Katalog »Serie 7« (Tecta) 5 Geflecht/Materialprobe »Serie 7« (Tecta),1977 6 Katalogseite »Serie 5«, 1975 (Tecta), 1977 7 Titelfoto Einzelmöbelprogramm »Art Deco«, 1979 (Reim Interline) 8 Franz Schulte, Mitarbeiter (1976-1978) 9 Laura Maly, meine Tochter, 3 Jahre alt 10 »Taxus« Solitärmöbelprogramm, 1977 (Reim Interline) 11 »Quadrat«, 1977 (Reim Interline) 12 Eibenholzfurnier »Taxus« 13 Skizzen »Art Deco«, 1979 14 Materialprobe, Geflecht von »Mutaro« 15 Titelfoto – erster Katalog »Mutaro«, 1979 16 Interieurfoto: Einrichtung des Hauses Lübke 17 »Mutaro«-Programm, 1979 (Interlübke) 18 Stuhl »Art Deco«, 1979 (Reim Interline) 19 Travertin-Füße »Taxus« 20 »Mutaro«-Materialien/Doppelseite Katalog, 1979 21 Ahornkegel (Unikat) 22 Griff »Art Deco« 23 Kommode »Serie 6«, 1976 (Tecta)

1 Photo of the »M 70« table, 1976 (Tecta) 2 »Tecta« catalogue title 3 Photo of Gisela Maly, stylist for the »Tecta« exhibit, 1975 4 Photo from the »Serie 7« catalogue (Tecta), 1977 5 Wicker/Material test for »Serie 7« (Tecta) 6 Page of the »Serie 5« catalogue, 1975 (Tecta) 7 Title photo for the »Art Deco« unique furniture program, 1979 (Reim Interline) 8 Staff member Franz Schulte (1976-1978) 9 My daughter Laura Maly at age three 10 »Taxus« unique furniture line, 1977 (Reim Interline) 11 »Quadrat«, 1977 (Reim Interline) 12 »Taxus« yew veneer 13 »Art Deco« sketches, 1979 14 Material test – »Mutaro« wicker surface 15 Title photo from the first »Mutaro« catalogue, 1979 16 Interior photo of the Lübke house 17 »Mutaro« program, 1979 (Interlübke) 18 »Art Deco« chair, 1979 (Reim Interline) 19 »Taxus« travertine base 20 »Mutaro« materials, double-page spread in catalogue, 1979 21 Maple cones (original) 22 »Art Deco« handle 23 »Serie 6« chest of drawers, 1976 (Tecta)

The Eighties Die achtziger Jahre

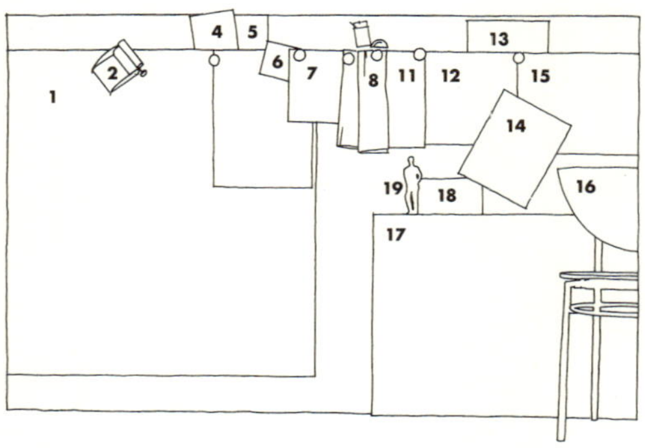

1 Zeichnung von »Zyklus« (1:1) als Poster – 1984/1994 (COR) 2 Modell 1:5 »Zyklus« 3 Entwurfsskizzen »Zyklus«, 1983/84 4 + 5 Fotos mit Mitarbeiter Hartmut Tromberend (1978-1987) 6 Polaroid »Zyklus«, Prototyp 7 Foto »Duo«-Schreibtisch, 1986 (Interlübke) 8 Textilien der Kollektion »Graphis«, 1985 (Interlübke) 9 Streichholzschachtel »Duo« 10 Griff »Duo« 11 Katalogseite »Duo-Bar«, 1986 12 Foto »Peter Maly Bett«, 1983 (Ligne Roset) 13 Foto mit Mitarbeiter Christian Werner (1987-1992) 14 Veröffentlichung vom Stuhl »Hera« (Ligne Roset) 15 Foto Recamiere »Antares«, 1988 (Ligne Roset) 16 Stuhl »Hera«, 1990 (Ligne Roset) 17 Isometrie Interlübke-Messestand, 1988 18 Skizze »Prao«, 1985 (Ligne Roset)

1 Drawing of »Zyklus« (1:1) as a poster – 1984/94 (COR) 2 »Zyklus« model on a scale of 1:5 3 »Zyklus« design sketches, 1983/84 4+5 Photos with staff member Hartmut Tromberend (1978-1987) 6 Polaroid shot of the »Zyklus« prototype 7 Photo of the »Duo« desk, 1986 (Interlübke) 8 Textiles from the »Graphis« collection, 1985 (Interlübke) 9 »Duo« matchbox 10 »Duo« handle 11 Catalogue page featuring the »Duo« bar, 1986 12 Photo of the »Peter Maly Bed«, 1983 (Ligne Roset) 13 Photograph with staff member Christian Werner (1987-1992) 14 Publication of the »Hera« chair, 1990 (Ligne Roset) 15 Photo of the »Antares« day bed, 1988 (Ligne Roset) 16 Chair »Hera«, 1990 (Ligne Roset) 17 Isometrics of the Interlübke trade fair stand, 1988 18 »Prao« sketch, 1985 (Ligne Roset)

The Nineties I Die neunziger Jahre I

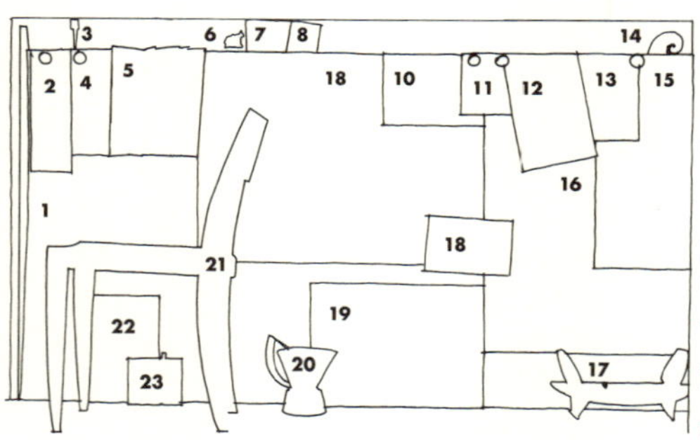

1+3 Griffe des »Metrix«-Programms, 1991 (Behr) 2 Einladung zur Premiere von »Metrix«, 1991 (Behr) 4+5 Katalog »Metrix«, 1991 (Behr) 6 Lineol-Tier – Rettungshund unseres Ateliers 7 Mitarbeiter Christian Werner (1987-1992) und Renate Dipper (1988-1990) 8 Mitarbeiterin Karin Görling (seit 1985) 9 Poster von einer Zeichnung der Liege »Cirrus« (Design von 1992) 10 Foto »Schöner Wohnen«, »Peter Maly Edition«, 1990/92 (Intra Design) 11 Foto der Präsentation der Teppich-Edition (JAB) 12 Foto der Bettwäsche-Kollektion »Margo«, 1994 (Domitex) 13 Foto vom Solitärschrank »Elliptic«, 1990 (Intra Design) 14 Metallintarsie aus der Teppich-Edition (JAB) 15+16 Collage und Poster der Teppich-Edition (JAB) 17 Modell M 1:5 der Liege »Cirrus«, 1992 (COR) 18 Skizze von »Sharky«, 1992 (Brunati) 19 Veröffentlichung von »Cycla« in »DDN«-Milano 20 Teekanne aus der Serie »Cycla«, 1993 (Wilkens) 21 Stuhl »Tectus«, 1991 (Bros's) 22 Titel des »Metrix«-Katalogs 1991 23 Erstes Modell von »Metrix«, 1990

1+3 Handles from the »Metrix« program, 1991 (Behr) 2 Invitation to the debut of »Metrix«, 1991 (Behr) 4+5 »Metrix« catalogue, 1991 (Behr) 6 Our studio mascot 7 Staff members Christian Werner (1987-1992) and Renate Dipper (1988-90) 8 Staff member Karin Görling (as of 1985) 9 Poster of a drawing of the »Cirrus« lounge chair (designed in 1992) 10 Photo of »Peter Maly Edition« in »Schöner Wohnen«, 1990/92 (Intra Design) 11 Photo of the presentation of the rug series (JAB) 12 Photo of the »Margo« bed linen collection, 1994 (Domitex) 13 Photo of the »Elliptic« unique cupboard, 1990 (Intra Design) 14 Metal inlay from the rug series (JAB) 15+16 Collage and poster of the rug series (JAB) 17 Model of the »Cirrus« lounge chair (1:5 scale), 1992 (COR) 18 Sketch of »Sharky«, 1992 (Brunati) 19 Publication of »Cycla« in »DDN« – Milano 20 Teapot from the »Cycla« series, 1993 (Wilkens) 21 »Tectus« chair, 1991 (Bros's) 22 Title of the »Metrix« catalogue, 1991 23 First »Metrix« model, 1990

The Nineties II Die neunziger Jahre II

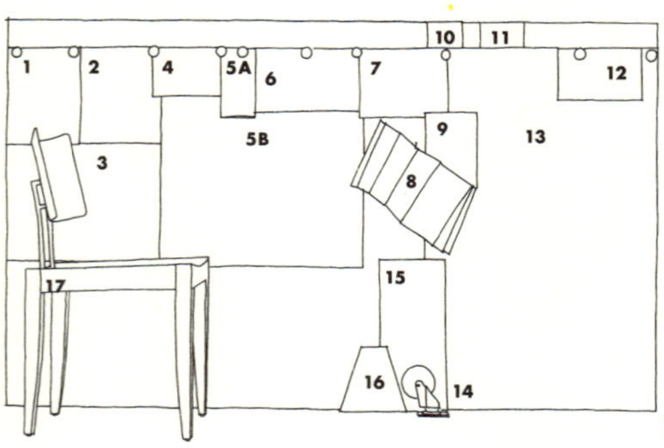

1 Katalog-Titel »Basic«, 1994 (Reim Interline) 2+3 Entwurfszeichnungen zu »Basic«, 1993 4 Foto der Kollektion »Parallele«, 1995 (Ligne Roset) 5A Einladungskarte mit Stuhl »Cena«, 1994 (Tonon) 5B Entwurfszeichnung »Cena«, 1994 6 Modell des »Ligne Roset«-Messestandes, IMM 1995 in Köln 7 Einladung zur »Thonet 737«-Präsentation, 1994 8 Präsentations-Folder »Thonet 737«, 1993 9 Foto »Thonet 737«, 1994 10 Foto des Teams anläßlich der »Thonet 737«-Präsentation in meinem Atelier 11 Das Team 1994/95 (von links: Karin Görling, Heike Endlich, Jens Ahlemeyer und Peter Maly) 12 Foto vom Sofa »Incognito«, 1995 (Ligne Roset) 13 Entwurfszeichnung 1:1 von »Ticino« (WK Wohnen) 14 Rolle des »Ticino«-Programms, 1994 15 Profilierte Alu-Fläche, Studie zur »Parallele« 16 Konstruktionsdetail des Stuhls »Cena« (Tonon): Aluplatte mit Prägung 17 Stuhl »Cena«, 1994 (Tonon)

1 »Basic« catalogue title, 1994 (Reim Interline) 2+3 Design sketches for »Basic«, 1993 4 Photo of the »Parallele« collection, 1995 (Ligne Roset) 5A Invitation featuring the »Cena« chair, 1994 (Tonon) 5B Design sketches for »Cena«, 1994 6 Model of the »Ligne Roset« trade fair stand, at IMM 1995, in Cologne 7 Invitation to the presentation of »Thonet 737«, 1994 8 Presentation pamphlet for »Thonet 737«, 1993 9 Photo of »Thonet 737«, 1994 10 Design team photo at the presentation of »Thonet 737« in my studio 11 The 1994-95 team (from left: Karin Görling, Heike Endlich, Jens Ahlemeyer and Peter Maly) 12 Photo of the »Incognito« sofa, 1995 (Ligne Roset) 13 Design sketch of »Ticino« (scale 1:1) (WK Wohnen) 14 Caster from the »Ticino« program, 1994 15 Profiled aluminium sheet, a study for »Parallele« 16 Construction detail of the »Cena« chair (Tonon): stamped aluminium plate 17 »Cena« chair, 1994 (Tonon)

Können Sie den Hintergrund Ihrer persönlichen Entwicklung beschreiben?
In den Beruf startete ich mit meiner Arbeit für die Redaktion »Schöner Wohnen« in den sechziger Jahren. Damals wurde sicher eine Grundlage gelegt, die mich auch in meiner späteren Tätigkeit als Designer nachhaltig beeinflußt hat.
Meine Tätigkeit innerhalb der Redaktion war die eines Innenarchitekten. Sie begann sehr profan mit der Beantwortung von Leserbriefen, in denen Leser ihre Wohnprobleme schilderten und um professionelle Unterstützung baten. Diese Hilfe bestand in Grundriß-Lösungen und perspektivischen Raumskizzen.
Viele solche zeichnerischen Lösungsansätze fanden Eingang in die Zeitschrift. Auch ich habe im Zuge dieser umfangreichen Service-Leistung eine Menge gelernt. Es war eine Art »Basisarbeit« für mich, die mir konkrete Wohnsituationen vor Augen führte und mich mit realistischen Problemen und Wünschen der Menschen bekannt machte.
Später inszenierte ich aktuelle Möbel und entwarf Räume für Fotoproduktionen. Es handelte sich um Bühnenbilder, künstlich erschaffene, ephemere Wohnwelten – nur für den Augenblick des Fotoshootings existent. Dennoch hatten sie stets eine reale Basis, bauten sie doch auf konkreten Grundrissen auf und verdeutlichten dem Leser räumliche Zusammenhänge.
Nirgends konnte man besser Erfahrung im Inszenieren von Räumen sammeln als in Fotostudios. Sofort war zu sehen, welche Fehler man gemacht hatte, und man konnte sie sogar noch korrigieren. Handwerker standen bereit, um jede Korrektur umgehend durchzuführen. War zum Beispiel die Decke zu dominant, wurde sie umgestrichen; war die Lichtstimmung nicht optimal, wurde neu ausgeleuchtet; alles ließ sich schnell verändern. Diese Erfahrungen stecken mir heute noch im Kopf, wenn ich Messestände inszeniere, Einrichtungen entwerfe.
Inwieweit kann man »Schöner Wohnen« als Medium sozialer Utopien oder auch als Machtinstrument sehen?
Man muß sich klar machen, daß »Schöner Wohnen« in den sechziger Jahren eine ganz andere Bedeutung hatte als heute. Heutzutage gibt es eine Reihe von Zeitschriften, die sich mit Innenarchitektur und Architektur beschäftigen. Damals aber hatte »Schöner Wohnen« eine Monopolstellung inne und daher eine sehr große Bedeutung als trendbildende Institution. Diese Verantwortung war der Redaktion sehr wohl bewußt. Manchmal war es geradezu beängstigend, wie man Trends machen konnte, indem man ein Wohnthema nach sehr persönlichen Auffassungen gestaltete. Umgehend fanden sich die eigenen Entwürfe multipliziert in Anzeigen unterschiedlichster Hersteller wieder, die auf diesen vermeintlichen Trend reagierten.
Wir waren eine sehr junge, engagierte Redaktion und hatten genaue Vorstellungen, wie zeitgemäßes Wohnen aussehen sollte; wir wollten unsere Utopie eines besseren Wohnens vermitteln, wollten dazu beitragen, Deutschland aus dem Abseits der Nachkriegszeit ans Licht zu holen. Stets vor Augen hatten wir das große Vorbild Skandinavien mit einer wesentlich höheren, eigenständigen Wohnkultur. Die deutsche »Wohnkultur« stand hierzu in krassem Gegensatz: unterentwickelt, außerordentlich fremd-orientiert und dadurch zwischen wechselnden Einflüssen hin- und hergerissen.
Unsere vornehmliche Zielsetzung war es daher, dem Leser zu vermitteln, daß es um den Ausdruck seiner ganz eigenen Persönlichkeit ging. Ihm sollten eine Reihe von Angeboten gemacht, Themen, die stilistisch unterschiedlich waren, vorgestellt werden. Beispielsweise wurde ein Raum in mehreren Varianten eingerichtet – eine Hilfestellung für den Leser, die persönliche Wohnform frei von überkommenen Wohnklischees zu finden.
Die F.A.Z. schrieb 1994 in einem Beitrag – wie ich finde sehr treffend: »Peter Maly kennt sich aus in deutschen Wohnzimmern.«
Wann haben Sie selbst angefangen zu entwerfen?
Schon während der Zeit bei »Schöner Wohnen« waren meine ersten Design- und Produktentwürfe entstanden – allesamt Ausdruck ihrer Zeit und als Antwort auf einen Mangel an Adäquatem gedacht. Diese ersten Möbel waren eine Reaktion auf die Ära der Pop-Art, auf die Zeit der Beatles, auf die Flower Power der Hippies und die Op-Art. Man arbeitete damals sehr experimentell, alle Konventionen wurden in Frage gestellt. Und so sahen die ersten Möbel, die ich damals entwarf, auch aus: außerordentlich farbenfroh lackiert, mit orange- und pinkfarbenen Dekoren, aber auch schwarz-weiß oder olivgrün mit violett – provokante Kombinationen, die ich zusammen mit dem Maler Michel Dupont entwickelte. Aber die heute exzentrisch wirkenden Farben wurden in den Sechzigern gar nicht als so schockierend empfunden, vielmehr waren die Objekte äußerst erfolgreich.
Diese ersten Möbel, die ich mit dem Hamburger Einrichter Rudolf Beckmann mehr oder weniger in Eigeninitiative herstellte und in Paris auf der Messe präsentierte, hatten einen beachtlichen Kommunikationserfolg. Viele internationale Zeitschriften nahmen das Thema damals auf. In den sechziger Jahren hatte ich auch meine erste Doppelseite in »Abitare« – und die Firma Tecta übernahm die Möbel in ihre Kollektion. Dieser Erfolg machte natürlich Mut weiterzuarbeiten.
Wodurch haben Sie eine besondere Prägung erfahren?
Schon während meines Studiums und auch am Anfang meiner Berufstätigkeit in den sechziger Jahren faszinierten mich die dänischen Architekten am stärksten. Ich nenne hier nur ein paar berühmte Namen: Wegener, Kjaerholm, Jacobsen und natürlich Verner Panton. Das dänische Design schien mir etwas sehr Handfestes und Gewachsenes zu sein, und auch die dänische Architektur mit den schönen Strukturen ihrer einfachen Häuser kam meinen Vorstellungen vom modernen Wohnen sehr entgegen.
Was heute Mailand ist, war seinerzeit für mich Kopenhagen – das Mekka des Designs schlechthin. Es war ein absolutes »Must«, regelmäßig dorthin zu reisen, um sich über die stilprägenden Entwürfe zu informieren.
Verner Panton hat mich damals mit seinen geometrisch geprägten Elementarformen und seinen klaren Farben ganz besonders fasziniert. Es waren seine Tütensessel, sein Kunststoff-Freischwinger – heute wieder bei Vitra in Produktion – und natürlich nicht zu vergessen: seine fantastischen Inszenierungen, die mich nachhaltig beeindruckten.
An dieser Stelle sei mir ein Wort des Dankes an Verner Panton gestattet, der freundlicherweise die Einleitung für dieses Buch verfaßt hat.
Haben sich auch Bewegungen anderer Länder in Ihrer Arbeit niedergeschlagen?
Durch meine Tätigkeit in der »Schöner Wohnen«-Redaktion kam ich sehr früh mit allen Trend-Informationen in Berührung. Italiens Aufbruch in den sechziger Jahren, den ich von Anfang an mitverfolgte, hat mich außerordentlich begeistert und mein Denken in dieser Richtung sehr stark beeinflußt. Die Fantasie, die aus dem Süden herüberschwappte, diese Welle völlig neuer Möglichkeiten, neuer Materialien, neuer Farben, neuer Anwendungsformen von Kunststoffen – das alles war äußerst aufregend. Allerdings war ich damals noch kein Designer, sondern Innenarchitekt – eine Art Publizist, der Design inszenierte.
Ohne Frage hat das italienische Design vermutlich jeden beeinflußt. All die großen Namen, die zu der Zeit zum ersten Mal auftauchten: Joe Colombo, Marco Zanuso, Mario Bellini, nicht zu vergessen Achille Castiglioni. Diese Gestalter waren so gut, daß man als junger Designer bewundernd vor ihren Entwürfen stand und wenig Raum für eine eigene Entwicklung sah. Auch heute geht es mir bisweilen noch so,

daß mich auf einer Messe erstklassige Dinge eher blockieren als anregen.

Wann rückte das Bauhaus in Ihr Bewußtsein?

Das war erst später, in den siebziger Jahren. Während des Studiums wurde es uns von unseren Dozenten in seiner Bedeutung überhaupt nicht nahegebracht. So war es für mich später eine eher autodidaktische Annäherung an dieses Thema. Es nahm mich dann allerdings so in Beschlag, daß die Prägung bis heute spürbar ist. Ich erinnere mich auch an intensive Gespräche mit Axel Bruchhäuser, dem Inhaber der Firma Tecta. Er war und ist ein glühender Verehrer der Bauhaus-Künstler und exzellenter Kenner dieser Zeit.
Mit den Tendenzen der dreißiger Jahre habe ich mich seinerzeit sehr eingehend beschäftigt und fast alles gelesen, was an Literatur darüber zu finden war. Ich war fasziniert von der sozial-utopischen Ausrichtung, die diesem Gedankengut innewohnte, und der absoluten Klarheit der Entwürfe, die daraus resultierten.
Allerdings brachte solche Begeisterung und starke Identifikation auch den Nachteil mit sich, daß ich keinen konkreten Ansatz für eigene Arbeiten fand. In Anbetracht der Qualität der Objekte schien es mir unmöglich, etwas zu gestalten, das daneben Bestand haben könnte.
In der Tat ist einem Mies van der Rohe-Stahlrohrstuhl nichts Vergleichbares entgegenzusetzen, und es gibt auch keinen besser gestalteten kubischen Sessel als den von Le Corbusier. Hier etwas hinzufügen zu wollen, wäre absolut lächerlich und anmaßend.

Sie betonen häufig die Bedeutung geometrischer Formen innerhalb des Entwurfsprozesses. Womit fing das an?

In den siebziger Jahren hat mich der geometrisch geprägte Jugendstil Hoffmanns sehr beeinflußt – Reisen nach Wien vertieften diese Eindrücke. Es war das Spiel mit den Quadraten, das mich faszinierte. Daraus entstanden Möbel wie der Tisch »M 70« für Tecta mit einem Metallguß-Gestell, das aus sich nach oben auflösenden Quadraten bestand. Aber auch die Ursprünge des Vitrinenmöbel-Systems »Quadrat« stammen aus dieser Zeit.
Auf die Spitze getrieben wurde dieses Prinzip der Anwendung von Quadraten in der Konstruktion der Treppenanlage im Hause Lübke. Hier waren es Gitterwerke aus Holz im Quadratraster, die sich zu einer Wendeltreppe formten.
In den Achtzigern begann meine große Liebe zu geometrischen Elementarformen. Zum Quadrat gesellten sich die Kreisformen, die bis heute meine Arbeit prägen. So besteht beispielsweise das Möbel »Zyklus« ausschließlich aus Formen, die diese Elemente variieren: Tangential angeschnittene Kreisformen, angeschnittene Ringformen, runde Scheiben sowie Zylinder bilden das Profil.
Das Collagieren dieser Elementarformen – anschneiden, überschneiden, addieren –, das ganze phantasievolle Spiel mit diesen klaren Formen war und ist auch heute noch außerordentlich spannend und fruchtbar. Diese logisch begründbare Formsprache hat mich aber auch davor bewahrt, mich in Irrationalitäten zu verlieren.
Die Entwicklung von Alchimia und Memphis zum Beispiel fällt ja in diese Zeit. Sie hat mich aber überhaupt nicht in meiner Arbeit tangiert.

Sie sprachen soeben Alchimia und Memphis an. Wie ist Ihre Einstellung zu diesen Strömungen heute?

Die Design-Auffassungen dieser Gruppen widersprachen so ziemlich allem, was für mich wichtig war, standen doch die Objekte der Memphis- und Alchimia-Ära in polarem Gegensatz zu meinen eigenen Arbeiten dieser Zeit: Sie waren für meinen Geschmack viel zu schrill und laut, viel zu vordergründig, auch zu egozentrisch. Eine Horrorvorstellung, mit einer Ansammlung dieser Objekte zu wohnen!
Sie sollten ja auch nicht den Menschen dienen; vielmehr bestand ihre primäre Aufgabe darin, Aufmerksamkeit auf sich zu ziehen, zu provozieren und – last not least – ihre Schöpfer zum Gesprächsthema in den Medien zu machen. Das zumindest ist gelungen. Auf der Strecke blieben allerdings Nutzen, Vernunft, Klarheit, Funktionalität und Logik.
Eine gigantische Sackgasse, in die sich eine ganze Designergeneration verirrte. Sie wurde verunsichert und war kaum noch fähig, sinnvolle Produkte zu entwerfen und in Zusammenarbeit mit der Industrie zu entwickeln.
Aber diese Auseinandersetzung, die die Gestalter des »Bel Design« mit den genannten Gruppen führten, ist heute nur noch kunsthistorisch interessant.
Fast alle Memphis-Jünger sind inzwischen reumütig zum Industrial Design zurückgekehrt. Michele De Lucchi zum Beispiel entwirft heute absolut funktionales Industrie-Design, und an die Stelle der Provokation ist seine Arbeit für die Deutsche Bank getreten. Lediglich Ettore Sottsass ist als eine Art gottähnliche Gestalt übriggeblieben – unbehelligt von öffentlicher Kritik.
Als im vergangenen Jahr in den Hamburger Deichtorhallen eine große Sottsass-Ausstellung stattfand, wurde Kritik nur hinter vorgehaltener Hand geäußert. Irgendwie erinnert mich das Ganze ein wenig an Andersens Märchen »Des Kaisers neue Kleider«: Wer die (nicht vorhandenen) Kleider des Königs nicht sah, geriet ja auch in den Verdacht der Inkompetenz ...

Was ist das wichtigste Kriterium für Ihre Arbeit heute?

Innerhalb meiner Arbeit ist heute Langlebigkeit mein größtes Anliegen. Produkte zu machen, die über längere Zeit Bedeutung haben, ist aus zweierlei Gründen wichtig. Der erste ist der ökologische Aspekt: Vor dem Hintergrund unserer immer knapper werdenden Ressourcen ist es ein Wahnsinn, kurzlebige Konsumgüter zu produzieren. Derart mit Rohstoffen umzugehen, ist unverantwortlich.

Die Ökologie ist ein sehr wichtiger Aspekt, deren Bedeutung in Zukunft noch zunehmen wird, aber bereits heute für viele kaufentscheidend ist. Wie können Sie sich als Designer für ökologische Aspekte einsetzen?

Der Einfluß, den ein Designer auf die Industrie hat, wird häufig überschätzt – das muß einmal ganz klar gesagt werden. Vielmehr ist man als Gestalter nur einer von vielen am runden Tisch, an dem das Produkt entsteht. Man ist hier Teil einer Gruppe von Menschen, die Entscheidungen treffen, und kann seine Sicht der Dinge allenfalls werbend einbringen. Die primäre Aufgabe des Designers besteht meiner Ansicht nach darin, dem Produkt eine Form zu geben, die sich nicht an schnell wechselnden Trends orientiert, sondern über längere Zeit Bedeutung hat. Das ist sicher für mich die wichtigste Anwendungsform von Ökologie im Design: langlebige Produkte zu gestalten, Produkte mit Vererbungsqualität. Sie sparen nicht nur die meisten Ressourcen, sie brauchen auch nicht recycelt zu werden, sie landen nicht auf der Müllhalde, sondern werden vielleicht an die nächste Generation weitergegeben.
Der zweite Aspekt für langlebiges Design ist ein psychologischer: Man sollte den Menschen Halt geben, sie nicht permanent mit neuen Trends überhäufen oder ihnen ständig neue Designrichtungen aufoktroyieren, die sie nur verunsichern. Wir brauchen kein Design, das bestenfalls zum Partygespräch taugt! Produkte modisch zu machen, sie einem viel zu schnellen Rhythmus zu unterwerfen, ist geradezu verantwortungslos! Eine modische Form wird innerhalb kürzester Zeit überholt sein und beim Konsumenten Überdruß und Verunsicherung hervorrufen. Gerade in unserer Zeit aber brauchen und wollen die Menschen Halt und Sicherheit, die Designer sollten sie hierbei unterstützen.

Ist Ihnen der schwierige Spagat zwischen »Aktualität« einerseits und dem Ziel »Langzeitdesign« andererseits immer gelungen?

Leider nicht immer. In der zweiten Hälfte der achtziger Jahre war ich, angeregt durch die Architektur-Biennale in Venedig »The present of the past«, sehr vom Wege abgekommen. In der dort präsentierten »strada novissima« proklamierten Architekturstars wie Venturi, Tigermann, Stern, Moore, Graves, Krier, Hollein und andere »The end of prohibition«. In diesem phantasievollen Aufbruch in das Zeitalter postmoderner Architektur sah auch ich für mich ganz neue Gestaltungsmöglichkeiten und griff tief ins Repertoire der klassischen Architektur. Säulenpaare, geteilte Giebelformen, Turmspitzen und Rundbögen transferierte ich ins Möbeldesign. Die ersten »Facades«-Schränke »Minos«, »Nike« und »Luna« von 1987 sind Beispiele hierfür, aber auch »Veritas« von 1988. Sehr bald aber mußte ich mir eingestehen, daß ich mich in einer Sackgasse befand – schnell eingeholt von einem Heer von Epigonen, die diese Architekturzitate heftig mißbrauchten. Ich hatte alle Hände voll zu tun, um aus dieser allgemein ausbrechenden Giebel- und Säulenpest wieder herauszukommen. So wurde zum Beispiel die »Facades«-Serie um neue, reduzierte kubische Formen erweitert, um die postmodernen Architekturformen in den Hintergrund zu drängen und schließlich ganz abzulösen.

Fast alle Ihre Entwürfe werden industriell hergestellt. Eher handwerklich geprägte Objekte sind die Ausnahme. Wie würden Sie den Prozeß der industriellen Realisation beschreiben?

Die Entwicklung von Industrie-Produkten ist ein oft langwieriger, Kraft und Ausdauer erfordernder Prozeß. Er bedarf vor allem der Fähigkeit zur Teamarbeit – sowohl mit den Fertigungsfachleuten der Hersteller wie mit den für das Marketing Verantwortlichen. Natürlich muß man in dieser Phase sein »Kind« vor dem Zugriff vieler »Mit-Väter« und vor gutmeinenden Vereinfachern schützen. Erst in dieser Phase entsteht aus einer Idee ein wirkliches Produkt, erst hier entsteht unter der Mitwirkung von Fachleuten Detailqualität, erst dann entsteht auch ein sinnvolles Preis-Leistungs-Verhältnis.

Welchen Stellenwert hat für Sie der »gesamtheitliche Prozeß« beim Design?

Für mich beginnt er schon in der Entwurfsphase. Ich stelle mir ein Produkt, das ich entwerfe, nicht von der Realität losgelöst als Einzelobjekt wie einen Kunstgegenstand im Museum – weißer Hintergrund, Präsentation auf einem Sockel – vor. Von Anfang an stellt sich für mich die Frage, wie sich dieses Möbel mit den anderen Komponenten des Raums verbindet. Hier bin ich also gleichzeitig Produktgestalter und Innenarchitekt. Allerdings soll nicht eine vorgeschriebene Raumlösung am Ende meiner Überlegung stehen. Indem ich mir Räume vorstelle, in denen Menschen unserer Zeit leben, versuche ich, die vielfältigen Bedürfnisse der Kunden besser vorauszudenken.

Der Prozeß setzt sich fort, wenn der Entwurf als Produkt realisiert werden soll. Meiner Überzeugung nach darf die Vision, die der Entwerfer am Anfang hat, nicht verloren gehen. Deshalb versuche ich, alle Beteiligten, die mit diesem Produkt zu tun haben, auf eine gemeinsame Linie einzuschwören. Das fängt mit der Geschäftsleitung an: Ich bemühe mich, Begeisterung und Motivation für das Projekt zu wecken und zu erhalten. Auch die technische Entwicklung muß in diesem Sinne erfolgen. Schließlich ist es an dem Vertrieb, mit Überzeugung die Kommunikation mit dem Handel zu führen. Dieser rote Faden kann an vielen Stellen abreißen und die Vision damit unterwegs verlorengehen. Daher rührt mein Engagement in jeder Stufe des Projekts.

Eine weitere Schwelle, die auf dem langen Weg bis zum zukünftigen Benutzer und Käufer überschritten werden muß, ist die richtige Präsentation des Möbels auf Messen. Wie sehen Sie den Stellenwert solcher Inszenierungen?

Der Entwurf von Messeständen ist immer wieder eine eminent wichtige Aufgabe für mich, aber auch eine besonders reizvolle! Hier kann ich meine architektonischen Ambitionen ausleben und gleichzeitig den neuen Produkten einen adäquaten Rahmen geben.

In diesem Zusammenhang möchte ich auf die entscheidende Rolle meiner langjährigen Stylistin, meiner Frau Gisela Maly, hinweisen. Ihr gelingt es stets aufs neue, die von mir konzipierten Messearchitekturen durch perfektes und hochprofessionelles Styling in kultivierte Lebensräume zu verwandeln. Sie versteht es, einer umfangreichen, vielgesichtigen Möbelkollektion durch ihre Arbeit – oder sollte man besser Kunst sagen – eine alles verbindende stilistische Linie zu geben.

Auf Messen tritt das Produkt zum ersten Mal an die Öffentlichkeit – und trifft auf die Fachhändler. Erst wenn es gelingt, diese von der Bedeutung des neuen Produkts zu überzeugen – und dazu gehört auch die Beflügelung ihrer Phantasie –, hat das Produkt eine reale Chance auf den übervollen Märkten.

Das Gleiche gilt auch für Fotorealisationen: Das fast immer auch von mir entwickelte Fotokonzept hat die Aufgabe, das Produkt richtig zu positionieren: Nur wenn das im Studio inszenierte Wohnumfeld stilistisch dem Produkt entspricht, können Menschen überzeugt werden.

Ich schildere Ihnen diesen Prozeß so ausführlich, weil die Erfahrung zeigt, daß nur bei Engagement auf allen Ebenen – vom Produktdesign über Innenarchitektur, Messestandgestaltung und Fotoinszenierung – der Erfolg eines Produkts wahrscheinlich wird.

Was ist das Geheimnis Ihres Erfolgs?

Auch ich kann nur Vermutungen anstellen. Ich arbeite für eine Reihe von Firmen, die sehr erfolgreich sind und sogar expandieren. Viele der von mir geschaffenen Produkte stehen im weltweiten Vertrieb; und nimmt man dies als Maßstab, bin ich wohl erfolgreich. Was aber das Geheimnis ist, kann ich selbst schwer beantworten. Vermutlich spielen mehrere Faktoren eine Rolle.

Zum einen ist da der ganzheitliche Aspekt in meiner Arbeit. Ich gehe wahrscheinlich umfassender vor als andere, entwerfe nicht nur ein losgelöstes Produkt, sondern denke von Anfang an in räumlichen Zusammenhängen und Benutzerkategorien. Zum anderen könnte eines der Geheimnisse sein, daß ich einen Instinkt dafür habe, was Menschen sich wünschen, ohne es explizit zu artikulieren. Ich habe wohl eine Antenne dafür entwickelt, wie Menschen wohnen möchten und welche Bedürfnisse sie in diesem Zusammenhang haben. Mit Marktanalysen ist das überhaupt nicht zu erfassen. Vielleicht sind Designer ja mit der nötigen künstlerischen Sensibilität ausgestattet, um Ansprüche der Menschen zu erspüren, die in deren Köpfen noch gar nicht existent sind. Diese Ansprüche gilt es, in hochwertige, unverwechselbare, aber auch industriell zu fertigende Produkte zu verwandeln.

Dazu gehört natürlich auch, den Weg zu bereiten und die Kommunikation zu den Menschen einzuleiten. Wenn dieses alles gelingt – und nicht nur einmal, sondern über einen langen Zeitraum hinweg – dann ist das vielleicht eine Erklärung für meinen Erfolg.

Can you describe the background of your personal development as a designer?

I got started in my profession in the Sixties, working for the magazine »Schöner Wohnen« – a popular magazine comparable to »Better Homes and Gardens«. The experience I acquired during this period formed a basis which had a lasting influence on my later design work as well.
My activities on the editorial staff were those of an interior designer. My initial tasks were very mundane, answering letters from readers in which they described the problems they were having with their homes, and asked for professional assistance. This assistance consisted of providing layouts and perspective drawings of rooms.
Many of these sketches were printed in the magazine. And I too learned a great deal from providing this comprehensive service. It was a kind of »grassroots activity« for me, in which I was confronted with concrete living situations and put in touch with people's real problems and desires.
Later, I created arrangements of contemporary furniture and designed rooms to be photographed. These were stage sets, artificially produced, ephemeral worlds of home interiors – existing only for the moment when the photos were shot. Nonetheless, they always had a basis in reality, since they built on concrete ground plans, and showed the reader what could be done with a given amount of space.
There was no better place for gaining experience in staging rooms than photo studios. You could immediately see where mistakes had been made, and even had time to correct them. Workers were always standing by to take care of any corrections that had to be made, on the spot. For example, if the ceiling was too dominant, it was repainted; if the lighting did not create the right atmosphere, it was redone; everything could be altered very rapidly. The experience I gained in those days is still in the back of my mind today when I design trade fair exhibits or interiors.

To what extent can »Schöner Wohnen« be regarded as a medium for social utopias or as an instrument of power?

First, let us bear in mind that »Schöner Wohnen« played a completely different role back in the Sixties than it does today. Today, there are several magazines in Germany that deal with interior design and architecture. In those days, however, »Schöner Wohnen« had a monopoly, so it was very important as a trend-setter. The editorial staff was well aware of this responsibility. At times it was actually quite frightening – the way we could set trends by working up an interior design theme according to our personal predilections. Then, in no time, our own designs would show up reproduced in ads for the widest variety of manufacturers who had responded to this purported trend.
We were a very young, highly motivated staff, and we had very clear ideas on how contemporary homes should look; we wanted to convey our utopian concept of better living; we wanted to help pull Germany out of the darkness of the postwar era and into the light. Scandinavia was a great role model for us, with its considerably more elevated, independent »home decor culture«, to which the German home decor culture stood in stark contrast: underdeveloped, highly oriented to foreign influences and thus pulled to and fro by changing influences.
As a result, our chief goal was to convey to readers that it was a matter of expressing one's very own personality. They were to be presented with a selection to choose from, themes that were different in terms of style. For example, a room might be furnished in several variations – to help readers in their search for a personal home environment that was free of conventional clichés.
In 1994, an article in the Frankfurter Allgemeine Zeitung newspaper put it this way – and quite aptly, I find: »Peter Maly knows his way around German livingrooms«.

When did you first begin to create your own designs?

My first design and product sketches came about during my time at »Schöner Wohnen« – all of them expressions of their time and intended as an answer to the lack of adequate conditions. These first pieces of furniture were a response to the Pop Art era, the Beatles, the Flower Power of the hippies and Op Art. Work in those days was of a highly experimental nature, calling all conventions into question. And this is how the first pieces of furniture I designed at that time looked – varnished in extremely bright colours, with orange and pink decorations, but also black-and-white or olive green with violet – provocative combinations which I developed with the help of painter Michel Dupont. However, the same colours that are considered eccentric today were not so shocking in the Sixties; on the contrary, these objects were a huge success.
I produced these first furniture designs together with the Hamburg interior decorator Rudolf Beckmann, more or less on our own initiative, and presented them at the trade fair in Paris. They became quite the talk of the town – at that time, the topic was being taken up by a number of international journals. Also during the Sixties, I had my first double-page spread in »Abitare«, and the Tecta company made the furniture part of their product range. Of course, this success inspired me to continue.

Which experiences have most served to give your work its own character?

During my studies, and when I first started in my profession in the Sixties, Danish architects interested me the most. Wegener, Kjaerholm, Jacobsen and of course Verner Panton are only a few of the renowned figures whom I could mention in this context. Danish design seemed to me to be something very solid and mature. Furthermore, the lovely structures which Danish architecture gave to simple houses were very much in line with my concept of modern living.
In those days, Copenhagen was to me what Milan is today – the very Mecca of design. Going there on a regular basis was an absolute must, to inform myself of the designs that were setting the course for an entire style.
In particular Verner Panton's work fascinated me, with its basic geometric forms and clear tones. His chair in the shape of an inverted cone, his plastic side chair – in production again at Vitra today – and, of course, his fantastic arrangements – left a lasting impression on me.
I would like to take this opportunity to extend a word of thanks to Verner Panton, who kindly agreed to write the introduction to this book.

Did movements taking place in other countries also find reflection in your work?

Working for »Schöner Wohnen« gave me first-hand knowledge of all the latest trend information. I was extremely enthusiastic about the upheaval in Italy during the Sixties, which I followed right from the beginning, and which had a strong influence on my thinking along these lines. The imaginative inventions flowing in from the Southern countries came in on a veritable wave of entirely new possibilities, new materials, new colours, new uses for plastics. It was all very exciting. At that time, however, I was not a designer, I was still an interior designer – a journalist of sorts who staged design.
There is no question that Italian design influenced just about everybody. So many great names appeared for the first time during this period: Joe Columbo, Marco Zanuso, Mario Bellini, not to forget Achille Castiglioni. These designers were so good that, being a young designer, I admired their work to a degree that left little room for my own developments. Even today it occasionally happens that first-rate objects at a fair are more of a creative block than an inspiration.

When did you first become aware of the Bauhaus?

That only happened later, in the Seventies. Our professors did not help us to appreciate it at all during our studies. So, everything I eventually learned about it was self-taught. However, once I began to learn about it, it took such a hold of me that the effects are evident in my work even today. I can also remember intense discussions with Axel Bruchhäuser, the proprietor of Tecta. He has always been a glowing enthusiast of the Bauhaus artists and is an expert on this period.

In those days, I investigated the trends of the Thirties quite exhaustively, reading almost all the available literature on the subject. I was fascinated by the orientation to social utopias inherent in these ideas, and the absolute clarity of the designs they yielded.

Nonetheless, there was a drawback to this enthusiasm and strong sense of identification: I could not find a concrete point of departure for my own work. In view of the quality of the objects, it seemed to me an impossible task to design something that could even hold a candle to them. The truth of the matter is that nothing can compare to a steel tube chair by Mies van der Rohe, and there is also no better-designed cube-shaped chair than Le Corbusier's. It would be ludicrous and arrogant to wish to improve on that.

You often stress the importance of geometric forms during the design process. Where did that begin?

During the Seventies, Hoffmann's geometric art nouveau style influenced me greatly, and my travels to Vienna deepened these impressions. It was the game with the squares which fascinated me. This gave rise to furniture such as the »M 70« table for Tecta with a cast metal frame, consisting of squares that seem to dissipate in an upward direction. But the »Quadrat« glass cabinet system also has its origins in this period.

This principle of using squares culminated in the staircase construction in Helmut Lübke's private residence. In this case it was latticeworks made of wood in a square grid pattern, which together formed a spiral staircase.

My great love affair with basic geometric forms began in the Eighties. Circles were added to the square; they are characteristic of my work even today. The »Zyklus« furniture, for example, consists exclusively of forms which are variations on these basic elements: circles with tangential cuts, ring forms with cuts, round discs and cylinders form the profile.

Combining these elementary forms into collages – cutting them, superimposing them, putting them together – the entire imaginative game with these clear forms has always been extraordinarily exciting and productive. This logically justifiable formal idiom has also prevented me from getting lost in irrational thinking.

Alchimia and Memphis emerged during this period, for instance. However, these trends had absolutely no effect on my work.

You just mentioned Alchimia and Memphis. What is your attitude toward these currents today?

The design conceptions of these groups contradicted pretty much everything that was important to me. Objects of the Memphis and Alchimia era were diametrically opposed to my own work of that time: they were much too shrill and loud for my taste, much too superficial and egocentric. The thought of living in the midst of these objects is a nightmare!

Nor were they intended to serve people – their primary purpose was to attract attention to themselves, to provoke, and not least of all, to make their creators into a media sensation. In that respect, at least, they were successful. However, features such as utility, practicality, clarity, functionality and logic were all left by the wayside.

It was a huge dead-end, and an entire generation of designers turned down it. They lost their self-confidence and were hardly capable any longer of designing useful products and developing them in co-operation with industry.

Today, however, the confrontation between designers of »Bel Design« and the above-mentioned groups is of interest only to art history.

Nearly all the Memphis disciples have returned, repentant, to industrial design. For example, Michele De Lucchi creates industrial designs that are completely functional, and his work for Deutsche Bank has taken the place of provocation ...

Only Ettore Sottsass has remained as a sort of god-like figure, unruffled by public criticism.

When a Sottsass exhibition took place in Hamburg last year, criticism was only offered with great hesitation. Somehow, the whole thing reminded me a bit of Hans Christian Andersen's fairy tale »The Emperor's New Clothes«. Everyone who could not see the (non-existent) clothing of the emperor was suspected of being incompetent ...

What is the most important criterion for your work today?

Longevity is my greatest concern in my work today. Creating products that will be meaningful over a longer period of time is important for two reasons. The first is the ecological aspect: in light of our ever-dwindling resources, it is absolutely insane to produce short-lived consumer goods. It is irresponsible to treat raw materials in that way .

Ecology is a crucial aspect which will become even more important in the future, but for many people it is already a factor in deciding whether or not to buy. How can you, as a designer, champion ecological aspects?

First of all, the fact of the matter is that the influence which a designer has on industry is often overestimated. The designer is really just one of many at the round table on which the product is created. You are one of the people who make decisions, and you can try to make a case for your own view of things. As I see it, the primary task of a designer is to give the product a form that is not oriented to rapidly changing trends, but which is of lasting importance. As far as I am concerned, that is no doubt the most important application of ecology in design: creating products that last, products that can be passed down from one generation to the next. Not only do such products contribute most to saving resources; they also don't need to be recycled, nor do they land on the rubbish heap, for they will perhaps be bequeathed to the next generation.

The second aspect of long-term design is psychological: we should give people something they can hold on to, instead of constantly bombarding them with new trends or forcing new design styles on them – which only makes them insecure. We do not need design which is, at best, a good conversation piece! Creating fashionable products and subjecting them to a rhythm that is much too fast-paced is downright irresponsible! A fashionable design will soon be out of style in the shortest time, and is ultimately bound to leave consumers fed up and insecure. Yet, what people really want and need, especially nowadays, is stability and security. And designers should support them in this respect.

Have you always been successful in the difficult task of achieving a design which is both »up-to-date« on the one hand and »long-term« on the other?

Not always, unfortunately. During the second half of the Eighties I strayed quite far from the path, as I was excited by the architectural Biennale in Venice, »The present of the past«. In the »strada novissima« presented there, star architects such as Venturi, Tigermann, Stern, Moore, Graves, Krier, Hollein and others proclaimed »the end of prohibition«. I, too, saw completely new possibilities for my own design work in this imaginative upheaval during the age of post-modern architecture, and I delved deeply into the

repertoire of classical architecture. I transferred paired columns, split gables, spires and round arches to furniture design. The first »Facades« cabinets – »Nike«, »Luna« and »Minos« from 1987 – are examples of this, as is »Veritas« from 1988.
Soon afterward, however, I had to admit to myself that I was in a cul-de-sac. A veritable army of imitators had caught up with me, and were seriously abusing these architectural references. It took all my energy to escape from this gable-and-column plague that was breaking out everywhere. This is why the »Facades« series, for example, was expanded by new, reduced cube forms in order to push post-modern architectural forms into the background and then completely replace them.

Nearly all of your designs are manufactured industrially. Hand-crafted objects are the exception in your work. How would you describe the industrial realization of a design?

Developing industrial products is often a lengthy process demanding a great deal of strength and stamina. In particular, team work is required – with the producer's technical experts as well as with those responsible for marketing. Of course, during this phase one must protect one's »baby« to some extent from the influences of many »co-fathers«, and from well-meant simplifications. Only in this phase does an idea give way to a real product; only here does the help of experts result in quality down to the last detail, and only then do you achieve a reasonable cost-benefit ratio.

What priority do you assign to the »holistic process« in your work?

For me, this process begins in the drafting phase. I imagine a product which I then draft, but not as an individual object isolated from reality, like an *objet d'art* in a museum – white background, mounted on a pedestal. Right from the beginning, I ask myself, what links this piece of furniture with the other components in the room? Thus, I am at once a product designer and interior designer. However, my thought process is not meant to culminate in a prescribed solution for the room. By imagining the rooms in which people live today, I attempt to anticipate the manifold needs of the customers.
The process continues when the design goes into production. In my opinion, a designer must never lose sight of his original vision. This is why I try to get everyone involved with the product in question to pull together and take a common tack. This starts at the management level: I endeavour to arouse and sustain motivation and enthusiasm for the product. The same has to apply to a product's technical development. Finally, it is up to the sales people to communicate with business partners in a convincing way. This thread may be lost or broken at many points along the way, and the vision along with it, which is why I am so committed to every stage of the project.

Being able to present the furniture in an inspiring way at trade fairs is a further crucial step on the long road to the future consumers and buyers. How do you rate the importance of such presentations?

Designing trade fair exhibits is always an eminently important task for me, and one which is also particularly challenging! It gives me an opportunity to give free rein to my architectural ambitions while I provide the appropriate setting for a new product.
In this context, I would like to point out the crucial role of my stylist of many years, my wife Gisela Maly. She is always successful at transforming the trade fair architecture which I have conceived into cultured living spaces via perfectly executed and highly professional styling. Through her work – or, I should say, art – she can give an extensive, multi-faceted line of furniture a stylistic orientation which integrates everything into a cohesive whole.
The product makes its first public appearance at trade fairs – where it also faces the critical eye of speciality dealers. Only when a new product can convince these people of its importance – and this includes sparking their imagination – does it have a true chance of making it in a glutted market.
The same applies to photographic realizations as well. The photo concept, which I also nearly always devise, must correctly position the product in its environment. People will only be convinced if the living environment in the studio is attuned to the product in terms of style.
I am describing this process to you in such detail because experience shows that only when one is involved at all levels – ranging from product design to interior design, from working up trade fair exhibits to photographic arrangements of a product – only then will a product have a good chance at success.

What is the secret of your success?

Even I can only guess. I work for a number of companies that have been highly successful and are even expanding. Many of my products are being sold throughout the world, and if one takes this as the measure of success, then I am indeed successful. However, it is hard to say what the secret is. There are no doubt a number of factors.
On the one hand, there is the holistic aspect of my work. I probably take more factors into account than others; I do not merely draft an isolated product, but rather think in terms of spatial contexts and user categories right from the beginning. On the other hand, my secret could be that I have an instinct for knowing what people want without their articulating it explicitly. In fact, I have developed feelers for discerning how people would like to live, and their needs in this context. It is something which cannot be determined through market analyses. Perhaps designers are equipped with the necessary artistic sensibilities for sensing demands and desires that people themselves are not even aware of yet. These demands must then be transformed into products that are high-quality and unique, but which can also be produced industrially. Of course, this also means paving the way for the product and opening up communication with consumers. If all this works out – and I do not mean just once, but over a long period of time – it may be one explanation for my success.

Das Interview wurde im April 1995 von Alex Buck in Hamburg geführt.

The interview was conducted by Alex Buck in Hamburg in April 1995.

Wolfgang Nagel

Auf der Suche nach der Einfachheit oder: Meister des Dialogs

Die neunziger Jahre scheinen als das Jahrzehnt der »Neuen Einfachheit« in die Stilgeschichte einzugehen. Zwar ist das ausgehende zwanzigste Jahrhundert von einem Stilpluralismus ohnegleichen geprägt, doch ist deutlich zu erkennen, daß sich die minimalistischen und puristischen Tendenzen verstärken, die mal mit den Trend-Etiketten »Simple Chic« und »Country Look«, mal als »Natürlichkeit« oder »Bescheidenheit« daherkommen. Waren die prosperierenden achtziger Jahre gekennzeichnet von Opulenz und Prächtigkeit, von formalen Experimenten und Grenzüberschreitungen – »Memphis« und die Postmoderne konnten exotische Blüten treiben –, schlägt das Pendel jetzt zurück: ein bekannter Effekt. Auf den in der Rückschau durchaus dekadent anmutenden Rausch der Farben und Formen folgt das Bedürfnis nach Neuorientierung; die Dekade des Überflusses mündet in eine Besinnung auf das Wesentliche. Auf einmal liegt der Reichtum in der Beschränkung. Zu oft haben die Menschen die Erfahrung gemacht, daß sich vordergründige Effekte schnell abnutzen. Die Reizüberflutung allenthalben nährt die Sehnsucht nach Stille und Geborgenheit, Trend-

In Search of Simplicity or: The Master of Dialogue

It seems as if the Nineties will go down in the history of style as the decade of the »New Simplicity«. Although the end of the 20th century has been shaped by an unparalleled stylistic pluralism, it is plain to see that minimalist and purist tendencies are indeed gaining strength, sometimes with trendy labels such as »Simple Chic« and »Country Look«, in other cases as »The Natural Look« or »Modesty«. Whereas the prosperous Eighties were characterized by opulence and splendour, by experimentation with forms and crossing boundaries – »Memphis« and post-modernism have spawned some exotic material – the pendulum is now swinging back in the other direction, a well-known effect. The intoxicating abundance of colours and forms, which in retrospect indeed seems decadent, is being followed by the need for reorientation; the decade of affluence is giving way to an era of getting back to the basics. Suddenly, wealth lies in setting limits. All too often, people have discovered that superficial effects quickly lose their appeal. The overstimulation surrounding us fuels the desire for peace and security; trendwatchers have coined the word »cocooning« to describe this phenomenon. And not least of all, more than ever before we are becoming aware of the fact that the finite resources of the earth and responsibility for the environment are forcing us to give a great deal of thought to the way we deal with materials. Self-respecting producers are voluntarily undergoing an eco-checkup, recycling is the order of the day. Against this background, the view of design has changed fundamentally.

Armlehnenstuhl »Thonet 737«, Thonet, 1994

»Thonet 737« armchair, Thonet, 1994

melder fanden dafür das Wort »cocooning«. Und nicht zuletzt wird uns stärker denn je bewußt, daß die endlichen Ressourcen der Erde und die Verantwortung für die Umwelt uns zu einem wohlbedachten Umgang mit Materialien zwingen. Hersteller, die auf sich halten, unterziehen sich einem freiwilligen Öko-Check, Recycelbarkeit ist ein Gebot der Stunde. Vor diesem Hintergrund hat sich der Blick auf das Design grundlegend geändert.
Neue Einfachheit – zweifellos gehört Peter Maly zu ihren Protagonisten. Mehr noch: Er ist einer ihrer Wegbereiter. Dabei hat der Hamburger Designer sich stets zugute gehalten, zeitgeistige Trendansagen zu ignorieren. »Schneller Wechsel ist der Tod jeder sinnvollen Entwicklung«, sagt Maly, »in unserer heutigen Zeit, in der das Konsumkarussell sich immer schneller dreht, ist es wichtiger denn je, den Menschen Halt und Perspektive zu geben. Das kann ich nur durch Kontinuität erreichen.« Gelassen und unbeirrbar hat er über Jahre hinweg an seiner klaren Formensprache festgehalten und kann heute mit Genugtuung beobachten, wie seine Positionen immer mehr Anhänger finden. Sein gegenwärtiger Erfolg hat denn auch nichts mit vermeintlichen Trends zu tun, aber sehr viel mit seinem Selbstverständnis als Designer.
Peter Maly fühlt sich dem Bauhaus verpflichtet und der Natur verbunden; er liebt die russischen Konstruktivisten und bewundert japanische Architektur. Er muß sich nicht auf Einfachheit zurückbesinnen, er hat vielmehr, von einem kurzen Flirt mit der Postmoderne (mit dem Programm »Facades« für Reim Interline) abgesehen, nie etwas

Von links nach rechts
Singulärschrank aus der »Facades«-Reihe »Cubic«, Reim Interline, 1988

Singulärschrank aus der »Facades«-Reihe »Quadro«, Reim Interline, 1992

Singulärschrank aus der »Facades«-Reihe »Undine«, Reim Interline, 1992

From left to right
»Cubic« unique cupboard from the »Facades« series, Reim Interline, 1988

»Quadro« unique cupboard from the »Facades« series, Reim Interline, 1992

»Undine« unique cupboard from the »Facades« series, Reim Interline, 1992

New Simplicity – Peter Maly is definitely one of its protagonists. In fact, he is one of its pioneers. And the Hamburg designer has always considered it to his credit that he ignores the trends promulgated by the Zeitgeist. »Rapid change is the death of any meaningful development,« Maly says. »Today, when the consumption merry-go-round is turning faster and faster, it is more important than ever to give people something they can hold onto, some sense of perspective. That is something I can only achieve through continuity.« Calmly and unerringly, he has held to his clear language of forms over the years and today can observe with satisfaction that his attitudes and convictions are finding increasing numbers of followers. And his current success has nothing to do with supposed trends, but much more with his understanding of his role as a designer.

Peter Maly feels indebted to the Bauhaus and to nature; he loves the Russian constructivists and admires Japanese architecture. He does not have to return to simplicity. On the contrary, aside from a brief flirt with post-modernism (the »Facades« series for Reim Interline), he never had anything other than simplicity in mind. As he

Vor meinem Vorstellungstermin bei Peter Maly rief ich zunächst eine befreundete Redakteurin an, kannte ich doch außer dem Sessel mit den Rädern nichts von ihm.

Angemessen aufgeregt betrat ich im Frühjahr 1987 das Hamburger Atelier mit einer großen Mappe voller Fotos, Zeichnungen, Skizzen von Studienarbeiten, Ideen und Entwürfen – insgesamt ein ziemliches Durcheinander. Peter Maly sah sich alles sehr aufmerksam an und kommentierte gnadenlos. Trotzdem bekam ich den Job.

Als er kurz darauf in Urlaub fuhr, übertrug er mir die Aufgabe, für einen Kunden etwas »ganz Neues« zu machen. Ich gab mir alle Mühe. Die Möbel gingen schließlich sogar in Produktion und wurden, wenn ich mich recht entsinne, ein beträchtlicher Flop.

Dennoch: Ich durfte bleiben. Ja, ich blieb viel länger, als ich es selbst ursprünglich geplant hatte. Mit gutem Grund. Ich schätzte das sehr direkte Arbeitsverhältnis des kleinen Büros ebenso wie den intensiven Gedankenaustausch zwischen Kollegen. Denn Peter Maly war mir bald mehr Freund als Chef, von dessen extrem disziplinierter Arbeitsweise ich außerordentlich profitiert habe. Bei ihm lernte ich weitaus mehr als einen geraden Strich aufs Papier zu bringen: vor allem Selbstkritik und Vorsicht vor allzu früher Euphorie.

Before my first meeting with Peter Maly, I rang up a friend of mine who was an editor to find out more about him: aside from the chair with the wheels I was not familiar with any of his work. Appropriately nervous, I entered the studio in Hamburg with a large portfolio of photographs, drawings, sketches of my work as a student as well as ideas and designs – all told, quite a mess. Peter Maly inspected everything very attentively and commented mercilessly. Despite this, I was offered the job.

Shortly after, as he was about to leave on holiday, Peter Maly instructed me to come up with something »completely new« for a customer. I gave it my best shot. The furniture even went into pro-

anderes im Sinn gehabt. Er will erklärtermaßen kein »Geschmacksdiktator« sein, der mit formalistischen Spielereien und vordergründigen Gags sein Publikum einzufangen sucht. Nach wie vor glaubt er an die Überzeugungskraft der Geometrie und die Notwendigkeit der Funktionalität.

Wo nicht länger aufgesetzte Schau-Effekte Aufmerksamkeit heischen, wo die Formen immer reduzierter werden, kommt es zwangsläufig auf andere Werte an, die in Zeiten der Originalitätssucht ins Hintertreffen geraten sind. Ausgesuchte, ökologisch »saubere« Materialien, hervorragende handwerkliche Verarbeitung, Raffinessen im Detail sind Komponenten eines neuen Qualitätsbegriffs. Die bewußten Verbraucher der neunziger Jahre sind kritischer, schauen genauer hin, lassen sich intensiver beraten – und sind auch bereit, für mehr Qualität mehr Geld zu bezahlen. Im Gegenzug erwarten sie allerdings, daß sie von ihren Möbeln länger etwas haben, nicht nur weil das Material, sondern auch die Ästhetik dauerhafter ist.

Für Peter Maly gehören die genannten Attribute seit jeher zu den Maximen eines verantwortungsvollen Designers. Als viele Verbraucher noch einer Ex-und hopp-Mentalität anhingen, hielt er am Ziel der Langlebigkeit fest. Darum kann er heute stolz auf mehrere Arbeiten verweisen, die seit mehr als zehn Jahren in achtbaren Stückzahlen

duction and, as I seem to recall, was a considerable flop. All the same: I was allowed to stay. Indeed, I stayed much longer than I had originally intended. With very good reason. I value the very direct working relations in a small studio, as well as the intensive exchange of ideas between colleagues that takes place there. For Peter Maly soon became more of a friend than a boss, and I have profited immensely from his extremely disciplined work mode. I learned far more from him than how to draw a straight line on a piece of paper: in particular, the ability to be critical of myself, and to beware of premature euphoria.

Christian Werner

Designer in Hamburg

Designer in Hamburg, Germany

himself has declared, he does not want to be a »dictator of taste« who tries to hook his audience with formalistic games and superficial gags. He still believes in the persuasive power of geometry and the need for functionality.
When contrived effects no longer upstage everything else, when forms become ever more reduced, other values necessarily come to the fore. These are values that had previously faded into the background during an era in which originality had become a sort of addiction: selected, ecologically »clean« materials, outstanding craftsmanship, refinement in detail are all components of a new concept of quality. The conscious consumers of the Nineties are more critical, look more closely, seek advice more intensively – and are also prepared to pay more money for better quality. In return, however, they expect to benefit longer from their furniture, because the aesthetics as well as the material are more capable of standing the test of time.
For Peter Maly, the above-mentioned attributes have always ranked among the maxims of a responsible designer. At a time when many consumers were still followers of a »throwaway« mentality, he held fast to the goal of creating long-lasting products. As a result, today he has several works to his credit which have been produced in respect-

Linke und rechte Seite
Einzelmöbel-Programm »Veritas«, Reim Interline, 1988

Left and right side
»Veritas« program of individual serially produced items, Reim Interline, 1988

Von oben nach unten
Tisch »Thor« aus der »Edition Peter Maly«, Intra Design, 1990

Barschrank »Elliptic« aus der »Edition Peter Maly«, Intra Design, 1990

Barschrank »Elliptic« geöffnet, Intra Design, 1990

From top to bottom
»Thor« table from the »Edition Peter Maly« series, Intra Design, 1990

»Elliptic« bar cupboard from the »Edition Peter Maly« series, Intra Design, 1990

Open view of the »Elliptic« bar cupboard, Intra Design, 1990

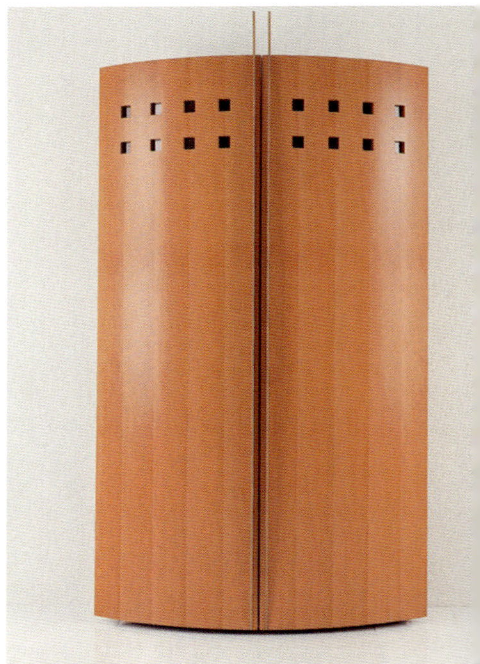

unverändert produziert werden, darunter das Sideboard- und Vitrinen-Programm »Quadrat« (1977, Reim Interline) oder der zweirädrige Sessel »Zyklus« (1984, COR), ein Vorläufer des in den Neunzigern propagierten »Mobilen Wohnens«.
Solch angehende Klassiker zu kreieren, ist allemal schwieriger, als mit einem vorgeblich originellen Entwurf einen Augenblicks-Erfolg zu erzielen. Für einen zeitlosen Gegenstand müssen von vornherein die Prioritäten entsprechend gesetzt sein. Maly macht die erste Haltbarkeitsprüfung bereits, bevor er zum Skizzenblock greift: Jeder neue Entwurf muß für ihn auf dem dicht besetzten Möbelmarkt eine eigene Existenzberechtigung haben. Was es bereits gibt, will er kein zweites Mal gestalten. »Ich möchte nicht die nächste Variante von irgend etwas machen«, sagt Maly und gesteht: »Vielleicht war ich früher, als ich anfing, nicht ganz so kritisch. Aber das ist schon lange her. Seit geraumer Zeit bin ich ziemlich streng mit mir und befasse mich nur mit Dingen, die ich selbst wirklich als sinnvoll ansehe.«
Grundvoraussetzung bei einer solchen Prüfung ist die Funktionalität – für Maly bleibt der Mensch das Maß aller Dinge, Design dient ihm als praktische Lebenshilfe. Sodann

Oben
»Edition Peter Maly«, Intra Design, 1990

Unten
Entwurf zum Einzelschrank »Papilio«, Intra Design, 1993

Above
»Edition Peter Maly«, Intra Design, 1990

Below
Draft of the »Papilio« cupboard, Intra Design, 1993

Linke Seite

Detail Schubladeneinsatz »Metrix«, Behr, 1992

Detail Treppe »Metrix«, Behr, 1992

Entwurf zum Schrankwand-System »Metrix«, Behr, 1992

Rechte Seite

Schrankwand-System »Metrix«, Behr, 1992

On the left

Detail of a drawer insert for the »Metrix« line, Behr, 1992

Detail of the »Metrix« bookcase stairs, Behr, 1992

Draft of the »Metrix« wall-to-wall cupboard system, Behr, 1992

On the right

»Metrix« wall-to-wall cupboard system, Behr, 1992

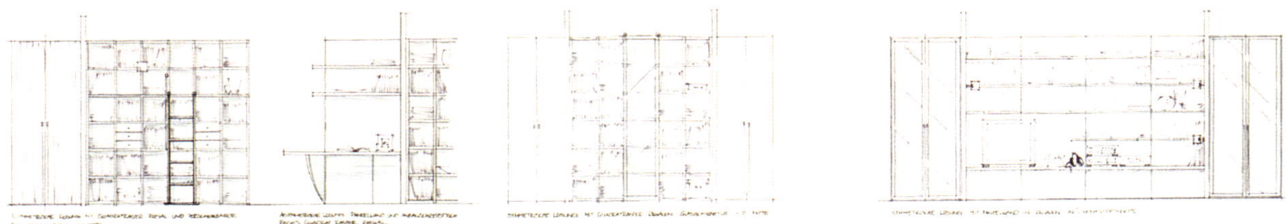

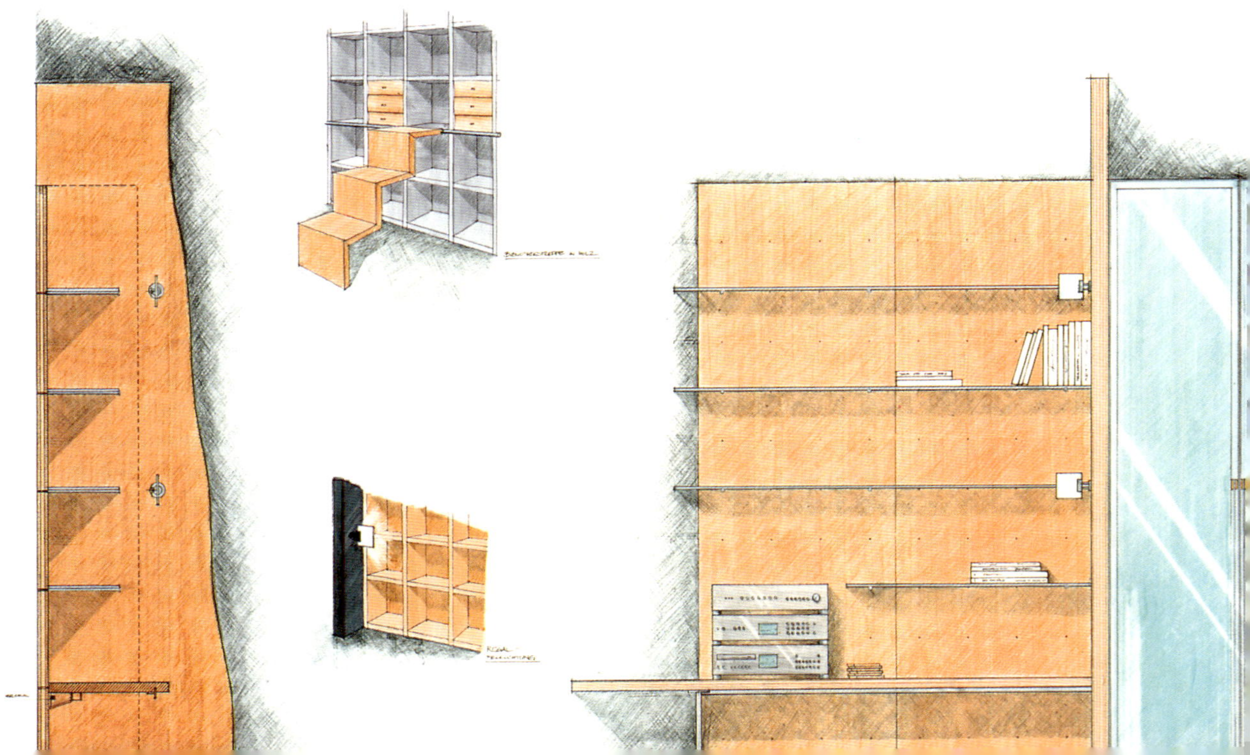

Über Begegnungen mit Peter Maly freut man sich immer, seien sie gewollt oder zufällig.
Sowohl im persönlichen Gespräch als auch in der zielgerichteten Arbeit überzeugt er durch seine Begeisterung und steckt mit ihr sowohl sein Team als auch seine Geschäftspartner an.
Dank seiner großen Erfahrung arbeitet er praxisnah und marktorientiert. Peter Maly ist von Anfang an – vom »briefing« über die ersten Ideen und Vorentwürfe – wirklich mit »feeling« dabei. Anschließend verfolgt er zielstrebig die Umsetzung von der Musterfertigung bis zur Serie. Der formale Gesamteindruck wird durch das Detail unterstützt. Dazu bewertet Peter Maly Materialeinsatz und Arbeitsgänge, seine Liebe zum Handwerklichen ist »greifbar«. Für eine preiswertere Produktion ist er aber auch bereit, sinnvolle Kompromisse zu schließen. So verwirklicht in »Metrix « – Wohnen und Schlafen – für Behr!

I always look forward to meetings with Peter Maly, whether planned or coincidental.
In personal conversation as well as in project work, his enthusiasm is contagious, and he is able to convince his team members and business partners alike.
His vast experience helps him to work with a market orientation, and in a way which is in keeping with actual practice. Right from the beginning – from the briefing on the project to the first ideas and preliminary drafts – Peter Maly is »all there« with a feeling for what is going on. Subsequently, he ambitiously pursues the realization of the project, from the production of the model to the actual series. The overall formal impression is enhanced by the numerous details. Furthermore, Peter Maly assesses the use of materials and production processes: his love of craftsmanship is »tangible«. At the same time, however, he is also willing to make sensible compromises for the sake of less costly production. Just as was realized in »Metrix« – quality living and sleeping by Behr!

Dr. Heinz Etzel

Geschäftsführer von Behr in Wendlingen

General Manager of Behr in Wendlingen, Germany

able quantities for over ten years, with no alterations. These include the »Quadrat« sideboard and glass cabinet series (1977, Reim Interline) or the two-wheel »Zyklus« easy chair (1984, COR), a precursor of the »Mobile Living« propagated in the Nineties.
Creating such prospective design classics is in any case more difficult than achieving short-term success with a supposedly original design. In order to create a timeless object, priorities must be established beforehand. Maly conducts an initial test of durability even before he picks up the sketch pad: Every new design which he sets out to draft must have a right to exist on its own terms in an already glutted furniture market. He has no intention of creating a variation on something that already exists. »I do not simply want to create the next version of a thing«, Maly says, and admits: »Years ago, perhaps, when I first started out, I was not quite as critical. But that was a long time ago. For quite some time now, I have been quite strict with myself, only dealing with things that I myself consider truly necessary.«
The basic prerequisite for passing such a test is functionality – in Maly's view, the human being is always the measure of all things, and design is intended to serve people

Mitte der siebziger Jahre arbeiteten wir erstmals im Rahmen einer Interlübke-Sonderausstellung auf der Messe zusammen. Dies machte uns soviel Spaß, daß die Geschäftsleitung der Meinung war, wir sollten Peter Maly mit der Entwicklung des Gesamtkonzepts für unsere Schaufenster betrauen.
Der Designer war einverstanden und brachte bereits in die Vordiskussion Aspekte ein, die in der Vergangenheit völlig unberücksichtigt geblieben waren. Dank Peter Maly drangen wir in eine Welt ein, die mit dem normalen Markt nichts gemein hatte. Wir wagten das Experiment und erzielten ein äußerst positives Ergebnis, das – wie sich später zeigte – auch kommerziell erfolgreich war. Seit über dreißig Jahren bin ich als Einkäufer und Beobachter auf allen wichtigen Messen Europas tätig. In dieser Eigenschaft habe ich Peter Maly als Entwerfer zahlreicher interessanter Produkte schätzen gelernt. Für mich ist er einer der ganz wenigen deutschen Designer, denen es gelungen ist, sich in der europäischen Designszene einen Namen zu machen. Einer meiner Favoriten ist die im letzten Jahr für WK Wohnen entwickelte »Ticino«-Serie.

In the mid-Seventies, we worked together for the first time on a special trade fair stand for Interlübke. It was so enjoyable that management decided to entrust Peter Maly with developing an overall design for our display windows.
He agreed, and, even during the preliminary discussions, brought up aspects which had previously been ignored altogether. Thanks to Peter Maly, we were able to break into a world which had nothing in common with the usual market. We embarked on an experiment which yielded highly favourable results, and later proved to be a commercial success as well.
For over 30 years, I have been active as a buyer and trendwatcher at all the important European trade fairs. Through these activities, I have come to admire Peter Maly as a designer who has created a great many interesting and unusual products. I consider him to be one of the very few German designers who has succeeded in making a name for himself on the European stage. One of my favourite Maly works is the »Ticino« series, designed for WK Wohnen in recent years.

Dieter Pesch

Mitglied des WK Wohnen-Aufsichtsrats und Geschäftsführender Gesellschafter des Einrichtungshauses Pesch Wohnen in Köln

Member of the WK Wohnen Supervisory Board and Managing Partner of the Pesch Wohnen furniture store in Cologne, Germany

muß ein neues Produkt eine formale Berechtigung haben, also »eine eigenständige Gestaltung mit kulturellem Anspruch« (Maly). Diese kann darin liegen, daß sie einen innovativen Weg weist, auf veränderte Wohngewohnheiten reagiert oder einer bestimmten Zeitstimmung adäquat Ausdruck verleiht.
Die Formgebung selbst ist bei Maly niemals ein willkürlicher Akt. Formen müssen für ihn logisch begründbar sein. Bei ihm haben Rundungen einen genau berechneten Radius, Proportionen orientieren sich am Goldenen Schnitt. Das klingt nüchterner als es aussieht, und oft genug hat Maly gezeigt, daß formale Strenge und Komfort, Rationalität und Sinnlichkeit einander nicht ausschließen. Unverdrossen glaubt er an den »Zauber der Geometrie«. »Die Geometrie«, sagt Maly, »hat mich immer fasziniert. Sie ist die Fundgrube für eine unendliche Vielfalt an Formen, mit denen ich spielen kann. Ich löse sie aus ihrer mathematischen Starrheit und füge sie neu zusammen zu phantastischen Collagen.« So gelangt er zu Schöpfungen voller Harmonie, die weitgehend gegen optischen Verschleiß resistent sind. Ein Sessel wie »Zyklus« leugnet nicht die Zeit seiner Entstehung, aber er ist heute noch genauso schön anzusehen wie vor einem Dutzend von Jahren.

as a practical aid in life. Therefore, it is only logical for a new product to have a formal justification – in other words, »an independent design with a claim to culture«, to quote Maly. This justification may consist in pointing the way in an innovative direction, responding to changes in living habits or giving adequate expression to a certain mood of the times.

Design itself is never an arbitrary act in Maly's case. In his view, forms must be logically justifiable. In his designs, curves have a precisely calculated radius, and proportions are oriented to the Golden Section. This sounds more serious than it actually is, and often enough Maly has indeed shown that formal stringency and comfort, rationality and sensuality, are not mutually exclusive. He is an undaunted believer in the »magic of geometry«. »Geometry«, Maly says, »has always fascinated me. It is the treasure trove for an infinite variety of forms with which I can play. I liberate them from their mathematical rigour and put them back together again in fantastic collages.« He thus achieves creations full of harmony which are largely resistant to optical wear and tear. An easy chair like »Zyklus« does not disavow the times in which it was created, but it is still just as pleasing to the eye today as it was a dozen years ago.

Von links nach rechts

Detail Schreibplatz »Metrix«, Behr, 1992

Detail Regalleiter »Metrix«, Behr, 1992

Detail ausschwenkbarer Bartisch »Metrix«, Behr, 1992

1. Arbeitsmodell aus Sperrholz und Pappe »Metrix«, Behr, 1991

From left to right

Detail of the »Metrix« writing table, Behr, 1992

Detail of the »Metrix« bookcase ladder, Behr, 1992

Detail of the »Metrix« swivelling bar table, Behr, 1992

»Metrix« plywood and cardboard mock-up, Behr, 1991

Linke Seite, von oben nach unten

Schranksystem für den Schlafraum »Metrix-Schlafen«, Behr, 1994

Innenansicht Schranksystem »Metrix-Schlafen«, Behr, 1994

Polsterbett und Lichtsäule »Metrix-Schlafen«, Behr, 1994

Paneelbett »Metrix-Schlafen«, Behr, 1994

Rechte Seite

Höhenverstellbarer Beistelltisch »Metrix-Schlafen«, Behr, 1994

Sprossenbett, Beistelltisch und Einzelkommode »Metrix-Schlafen«, Behr, 1994

On the left, from top to bottom

»Metrix-Schlafen« bedroom cupboard system, Behr, 1994

Interior view of the »Metrix-Schlafen« cupboard system, Behr, 1994

»Metrix-Schlafen« upholstered bed and pole-based luminaire, Behr, 1994

»Metrix-Schlafen« panel bed, Behr, 1994

On the right

»Metrix-Schlafen« side table with adjustable height, Behr, 1994

»Metrix-Schlafen« bed with slatted frame, side table and chest of drawers, Behr, 1994

Die Zusammenarbeit mit Peter Maly ist – heute wie auch vor achtzehn Jahren – wunderbar. Er hat fachlich und auch menschlich einen enormen Anteil an der Entwicklung unserer Kollektion. Als er vor fast zwanzig Jahren die Leuchte »MY« entwarf, verzweifelte er nicht an unserer Unerfahrenheit, vielmehr hat er uns mit großer Geduld und Sachlichkeit weitergeholfen.

Working with Peter Maly is wonderful – just as it was 18 years ago. He has made an enormous contribution to the development of our product line, not only professionally but also on a personal level. When he designed the »MY« luminaire nearly 20 years ago, our lack of experience did not cause him to despair; rather, he helped us immensely with his extraordinary patience and professional objectivity.

Gertrud Kracht

Inhaberin und Geschäftsführerin von Anta-Leuchten in Schenefeld

Owner and General Manager of Anta-Leuchten in Schenefeld, Germany

Auch bei der Materialwahl läßt sich Maly vom Gedanken der Langlebigkeit leiten. »Möbel sind aus Rohstoffen gefertigt, die viel zu wertvoll sind, um sie in immer schnellerem Rhythmus zu entsorgen oder bestenfalls zu recyceln.«
Sein liebster Werkstoff neben dem Metall war schon immer das Holz – schließlich absolvierte er vor der Ausbildung zum Designer eine Tischlerlehre. Seit Mitte der achtziger Jahre setzt er Holz verstärkt ein. Aber selten verwendet er es allein, vielmehr liebt er den Gedanken des Dialogs. Seine Entwürfe leben von Materialkontrasten. Maly nennt es »Dualität«. So wie er in seiner Formensprache Bogen gegen Gerade, Quadrat gegen Kreis setzt, verbindet er Materialien, die scheinbar gegensätzlich sind. Bei dem Sideboard-Klassiker »Quadrat« und dem Systemmöbelprogramm »Metrix« (1991, Behr) kombiniert er helles Kirschbaumholz mit schwarzen Lackflächen. Bei dem formal äußerst reduzierten Regalsystem »Basic« (1995, Reim Interline) erzeugen kühles Aluminium (für die Paneele) und warmes Kirschholz (für Fachböden und Fronten) eine reizvolle Spannung.
Die Stuhlkollektion »Cena« für Tonon (1995) behandelt die Materialunterschiede geradezu spielerisch: Auf ein unveränderbares Untergestell aus Buchenholz lassen sich

Maly also allows himself to be guided by the concept of longevity in his choice of material. »Furniture is made of raw materials which are much too precious to be disposed of at an increasingly rapid pace, or at best recycled.«

His favourite working material aside from metal has always been wood – after all, he was an apprenticed cabinetmaker before he was trained as a designer. He has been using wood to an increasing degree since the mid-Eighties. But he seldom uses it on its own, for he loves the idea of dialogue. His designs draw their vitality from material contrasts, which Maly refers to as »duality.« Just as he juxtaposes curves with straight lines, squares with circles, in his formal idiom, he also combines materials which are seemingly at odds with one another. In the sideboard classic »Quadrat«, and the »Metrix« system furniture series (1991, Behr), he combines light cherry wood with black lacquered surfaces. In the »Basic« shelf system, with its extremely reduced form (1995, Reim Interline), cool aluminium (for the panels) and warm cherry wood (for the shelves and front) create an intriguing tension.

The »Cena« chair collection for Tonon (1995) even makes a game out of the differences in material. Chair backs of various colours, forms and materials – wicker, alu-

Von links nach rechts

Tisch-Accessoires »Cycla«, Bremer Silberwaren (vormals Wilkens), 1993

Leseleuchte »Bibo«, Anta, 1991

Wandleuchte »Ludwigsburg«, Anta, 1994

From left to right

»Cycla« table accessories, Bremer Silberwaren (previously Wilkens), 1993

»Bibo« reading lamp, Anta, 1991

»Ludwigsburg« wall-mounted lamp, Anta, 1994

Linke Seite

Regalsystem mit Schiebetüren »Basic«, Reim Interline, 1994

Kommoden zu »Basic«, Reim Interline, 1994

Rechte Seite

Eßtisch »Schach« und Stuhl »Tectus«, Bros's, 1991

On the left

»Basic« shelf system with sliding doors, Reim Interline, 1994

Chests of drawers for the »Basic« system, Reim Interline, 1994

On the right

»Schach« dining table and »Tectus« chair, Bros's, 1991

minium, fabric – can be mounted on one and the same basic beechwood frame. The variability makes for seating arrangements that are individual and unconstrained, employing the widest variety of chairs – all of which, however, reveal the same common origin – much like family members sitting around a table.

Distinctive and exquisite – such were the combinations realized by Maly in the carpets for JAB Anstoetz. Here, he plays his geometric game with soft pile, smooth leather and shiny metal.

Recently, Maly has been falling back on natural materials to an increasing degree. Take linoleum, for example – this cheap floor covering had actually already been dropped from product lines, but it is made from recyclable raw materials (primarily linseed oil, natural resins and ground cork). It is enjoying a comeback on his Thonet table »Thonet 1737« (1995), where it provides a haptic contrast to the wood and intensifies the minimalist impression. The »Ticino« dining-room line (1995, WK Wohnen) also combines solid beechwood with linoleum inlays in friendly colours and woven baskets – Maly brings all the components together in a light, Mediterranean blend. For Ligne Roset, he designed the »Les points« carpet series from woven sisal, which is as simple as it is decorative. The only adornments are eyelets made of fine steel, which sprout tufts of cotton – if desired.

What Peter Maly likes most about natural materials is that they become more beautiful with use. Traces of use give them that precious patina which we so admire in antiques – because it can be acquired only through the natural ageing process. This is why Maly set out to design furniture which can be passed on from one generation to another: »Bequeathal rather than disposal is still the most environmentally sound way of dealing with materials.«

Ich begegnete Peter Maly vor zwölf Jahren bei der ersten Präsentation seiner Möbelentwürfe für Ligne Roset in Nizza. Damals waren ausschließlich südeuropäische Kunden eingeladen, und so herrschte fröhliches, lautes Stimmengewirr bei hochsommerlichen Temperaturen. Peter Malys farbenfrohe Sessel standen verstreut um den Swimmingpool und mittendrin hob sich der Designer in komplett schwarzem Outfit ab.

Im Frühjahr dieses Jahres kreuzten sich unsere Wege im Zuge der Vorstellung seiner ersten Entwürfe für Thonet erneut.

Dazwischen liegen zwölf Jahre kreativen Schaffens von Peter Maly. Ich schätze an ihm besonders, daß er sich als vielbeschäftigter Designer trotz seines großen Erfolgs eine stille Zurückhaltung bewahrt hat. Als Möbelentwerfer ist er zudem ein hervorragender Botschafter unserer Stadt.

I first met Peter Maly 12 years ago at the launch of his furniture designs for Ligne Roset in Nice. Only customers from Southern Europe had been invited to the event, and the torrid summer air was buzzing with a cheerful, loud babble of voices. Peter Maly's brightly coloured armchairs were strewn around the swimming pool, and the designer stood out among them, dressed in black from head to toe.

This past spring, our paths crossed once again during the presentation of his first designs for Thonet. And in the intervening 12 years, Peter Maly has been busily creative. What I especially admire about this designer is that he has remained quietly reserved despite his great success. Furthermore, as a furniture designer, he is an excellent representative of our city.

Alexander Raab

Inhaber und Geschäftsführer des Einrichtungshauses Clic in Hamburg

Owner and General Manager of the Clic furniture store in Hamburg, Germany

Rückenlehnen in verschiedenen Farben, Formen und Materialien – Korbgeflecht, Aluminium, Stoff – aufstecken. Diese Variabilität ermöglicht ebenso individuelle wie zwanglose Sitzrunden mit den unterschiedlichsten Stühlen, denen gleichwohl die gemeinsame Herkunft anzumerken ist – vergleichbar Verwandten, die sich um einen Tisch versammeln.
Ungewöhnlich und apart die Kombination, die Maly bei den Teppichen für JAB Anstoetz realisierte. Hier treibt er sein geometrisches Spiel mit weichem Flor, glattem Leder und glänzendem Metall.
In jüngster Zeit greift Maly immer stärker auf Naturmaterialien zurück. Linoleum, der eigentlich schon ausgemusterte Billig-Fußbodenbelag, der aber aus recycelbaren Rohstoffen (vorwiegend Leinöl, Naturharzen und Korkmehl) besteht, feiert auf seinem Thonet-Tisch »Thonet 1737« (1995) ein Comeback, sorgt dort für einen haptischen Kontrast zum Holz und steigert die minimalistische Anmutung. Auch das Eßzimmerprogramm »Ticino« (1995, WK Wohnen) kombiniert massives Buchenholz mit Füllungen aus Linoleum in freundlichen Farben und Körben aus Weidengeflecht – Maly führt alle Komponenten zu einer leichten, mediterranen Mischung zusammen. Für Ligne Roset entwarf er die ebenso simple wie dekorative Teppichserie »Les points« aus Sisalgewebe; das einzige Schmuck-Element sind Edelstahl-Ösen, aus denen – nach Belieben – Baumwollbüschel herauswachsen können.

Die Handschrift Peter Malys ist unverwechselbar und in fast jedem seiner Stücke wiederzuerkennen. Ohne sich zu wiederholen, schafft er es, Möbel neu zu interpretieren und den Entwürfen seine ganz besondere Aussage zu verleihen. Malys Gespür, wie Dinge aussehen sollen, und seine Liebe für Details machen mir bei der Umsetzung unserer Projekte besonderen Spaß! Seine ruhige, freundliche Art täuscht leicht darüber hinweg, daß sie auch einer fordernden Beharrlichkeit weichen kann. Eine Beharrlichkeit, die sich im nachhinein oft als richtig erwiesen hat.

Peter Maly's artistic signature is unmistakable, and can be recognised in nearly all of his works. Without repeating himself, he always manages to interpret furniture in an original way and lend a very special statement to his designs. Maly's sense of how things are supposed to look, and his love of detail, render the realization of our projects into an experience which I find particularly enjoyable! His quiet, friendly manner hides the fact that it can easily give way to a demanding persistence – which nonetheless often proves to be justified in retrospect.

Philipp Thonet

Mitinhaber und Gesellschafter von Thonet in Frankenberg

Joint-owner and partner of Thonet in Frankenberg, Germany

In the age of computers and mass production, we have a special appreciation for top-quality craftsmanship. Our eye is being attuned to elegant details and subtle refinements. Some of these refinements only take shape in the intensive dialogue with the craftsmen of the respective producers. This, too, is a permanent part of Maly's understanding of his role. As part of the difficult process of industrialization through which an idea becomes a product, he seeks direct contact as soon as possible with everyone involved, with workers as well as managers and marketing experts. »I try to convince everyone that we are not working on my product, but on a joint product. My experience has shown that if this process is not successful, one might as well break it off. And then, in most cases, the product is doomed to failure.«

As Maly knows very well, this »democratic« process means that everyone in the team must be prepared to make compromises – including the designer himself. Unlike some of his colleagues, however, he is not afraid that this will detract from his designs. Instead, he sees it as an opportunity to enhance quality. Disagreements with workers are tough at times. The designer may sometimes have to hear a craftsman explain that a joint will simply not hold up – and that a load test confirms this prognosis. The end result, however, is a higher-quality product. In his »Thonet 737« chair (1995, Thonet), for example, he had initially joined the wooden frame of the chair back with a mitre cut. After braiding it, he found he did not like the diagonal joint any longer. A cabinetmaker suggested that he should mortise the woods, and Maly immediately agreed – and was pleased at the added ornamentation effect created by the interplay of the different wood types. Such refinements in detail make an enormous contribution to the effect of what is actually a simple piece of furniture.

Von links nach rechts
Stuhlserie »Thonet 737«, Thonet, 1994

Eßzimmer mit Stühlen »Thonet 737« und Eßtisch »Thonet 1737«, Thonet, 1994

From left to right
»Thonet 737« chair series, Thonet, 1994

Dining room with »Thonet 737« chairs and »Thonet 1737« dining table, Thonet, 1994

An den Naturmaterialien schätzt Peter Maly am meisten, daß sie durch Benutzung schöner werden. Gebrauchsspuren verleihen ihnen jene kostbare Patina, die wir an Antiquitäten so bewundern – weil sie sich anders als durch natürliches Altern nicht herstellen läßt. Darum hat sich Maly vorgenommen, Möbel zu gestalten, die von einer Generation an die nächste übergeben werden: »Das Vererben, nicht das Entsorgen ist immer noch die umweltverträglichste Art, mit Materialien umzugehen.«
Im Zeitalter der Computer und der Massenfertigung wissen wir Spitzenerzeugnisse der Handwerkskunst wieder besonders zu schätzen. Unser Blick wird geschärft für elegante Details und kleine Raffinessen. Manche dieser Verfeinerungen entstehen erst im intensiven Dialog mit den Technikern des jeweiligen Herstellers. Auch dies ist ein fester Bestandteil des Malyschen Selbstverständnisses. In dem diffizilen Prozeß der Industrialisierung, in dem aus einer Idee ein Produkt wird, sucht er möglichst früh-zeitig den direkten Kontakt mit allen Beteiligten, mit den Handwerkern genauso wie mit den Managern und Marketingexperten. »Ich versuche alle davon zu überzeugen,

Von links nach rechts
Runder Eßtisch und Armlehnenstuhl »Ticino«, WK Wohnen, 1994

Vitrinenschrank »Ticino«, WK Wohnen, 1994

Detail Anrichte »Ticino«, WK Wohnen, 1994

Eßzimmer-Kollektion »Ticino«, WK Wohnen, 1994

Entwurf zu »Ticino«, WK Wohnen, 1994

From left to right
»Ticino« round dining table and chair with armrests, WK Wohnen, 1994

»Ticino« glass cabinet, WK Wohnen, 1994

Detail of the »Ticino« sideboard, WK Wohnen, 1994

»Ticino« dining room line, WK Wohnen, 1994

Draft of »Ticino«, WK Wohnen, 1994

Simplicity is often the result of a complicated process. There is no better proof of this than the »Thonet 737« chair. It is nothing less than the quintessence of Maly's creative work. In order to realize it, he developed a special relationship to it. He established contact with the Thonet company and consciously embraced the tradition of Mart Stam, Marcel Breuer et al. Measuring up to the masters of the Bauhaus over the decades may have been a source of inspiration for his work. In any case, the resulting chair possessed an elegant lightness reminiscent of the »Superleggera« by Gio Ponti or of the austere minimalism of Jasper Morrison's »Plywood Chair«. Nonetheless, it is a design of great originality. The chair prototypically unites what always distinguishes Maly's best designs – the love of geometry, the striving for reduction, the striking use of various materials, carefully reworked. With the production of a new, dirt-resistant cotton fabric for the woven surfaces, the chair also meets the criterion of innovation and is astonishingly comfortable.

It took Maly 36 months to design this chair, with a great deal of haggling over many of the fine points and details. Ultimately, however, it is the result of 30 years of creative activity and yet is entirely a product of its time, an icon of the Nineties. This is what constitutes its quality, its success.

The road to simplicity is far from over for Peter Maly. He is still searching. He still has the feeling of being able to find something new in playing with archetypal forms – the square, triangle and circle. This is perhaps one of the most surprising results of the »New Simplicity« – every successful new design demonstrates the extent to which the domain of the »simple« – supposedly so transparent – is still generating undreamed of variations, in spite of the fact that so many had already wanted to write it off as being »depleted«. »The opportunities«, as Peter Maly states confidently, »are far from exhausted.«

Unsere jahrelange Zusammenarbeit mit Peter Maly hat mit der Modellentwicklung »Ticino« einen Höhepunkt erreicht, mit dem sowohl der Zeitgeist als auch der sich ändernde Trend der Eßgewohnheiten erfaßt und in einer Kollektion umgesetzt wurde, die in Deutschland und Italien produziert wird.
Peter Maly ist für mich einer der wenigen Designer, die eine innere Mission konsequent verfolgen und daher eine unverwechselbare Handschrift besitzen. Es ist eine Freude, mit ihm in einem Entwicklungsteam zu optimalen Lösungen mit Kundennutzen zu kommen.

Our long-standing co-operation with Peter Maly reached a high point during the development of the »Ticino« model. It was a barometer for the Zeitgeist as well as the evolving trends in eating habits, and was translated into a line of furniture that is produced in Germany and Italy.
I consider Peter Maly to be one of the few designers who resolutely follow an inner mission, and who as a result possess an unmistakable signature. It is a pleasure to work with him on a development team to find optimal solutions that are of benefit to customers.

Hans-Herdin Duden

Geschäftsführender Gesellschafter von WK Wohnen in Leinfelden-Echterdingen

President and partner of WK Wohnen in Leinfelden-Echterdingen, Germany

daß wir nicht an meinem, sondern an einem gemeinsamen Produkt arbeiten. Wenn dieser Prozeß nicht gelingt, kann man nach meiner Erfahrung eigentlich abbrechen. Dann ist ein Produkt in den meisten Fällen zum Scheitern verurteilt.«

Maly weiß genau: In diesem »demokratischen« Prozeß muß jeder im Team zu Kompromissen bereit sein – auch der Designer selbst. Im Gegensatz zu manchem Kollegen fürchtet er jedoch nicht, daß sein Entwurf dadurch verwässert wird, sondern er sieht darin die Chance einer Qualitätssteigerung. Die Auseinandersetzungen mit Handwerkern sind manchmal hart. Da kommt es schon vor, daß sich der Designer von einem Techniker sagen lassen muß, daß eine Verbindung nicht ausreichend haltbar ist – und daß ein Belastungstest diese Prognose bestätigt. Aber am Ende steht meist ein besseres Ergebnis. Bei seinem Stuhl »Thonet 737« (1995, Thonet) zum Beispiel hatte er den Holzrahmen der Rückenlehne zunächst mit einer Gehrung zusammengefügt. Nach dem Beflechten gefiel ihm die diagonale Verbindung nicht mehr. Ein Tischler schlug vor, die Hölzer zu verzapfen, worauf Maly sofort einging – und sich freute, daß durch das Ineinandergreifen der Hölzer zudem ein Ornament entstanden war. Solche Verfeinerungen im Detail tragen enorm zur Wirkung eines eigentlich schlichten Möbels bei. Hinter der Einfachheit steckt oft ein komplizierter Prozeß. Dafür gibt es keinen im Wortsinn schöneren Beweis als eben diesen Stuhl »Thonet 737«. Er ist nichts Geringeres als die Quintessenz von Malys Schaffen. Um ihn realisieren zu können, ist er eine Wahlverwandtschaft eingegangen. Er hat den Kontakt zur Firma Thonet gesucht, um sich bewußt in die Traditionsreihe von Mart Stam, Marcel Breuer & Co einzugliedern.

Linke Seite
Stuhl- und Tischfamilie »Cena«, Tonon, 1995

Entwurf zu »Cena«, Tonon, 1995

Prototyp »Cena«, Tonon, 1995

Rechte Seite
Bett auf Rollen »Parallele«, Ligne Roset, 1995

Entwurf zu »Habitus«, Ligne Roset, 1995

Eßtisch »Biblia«, Ligne Roset, 1993 und Stuhl »Hera«, Ligne Roset, 1990

On the left
»Cena« table and chair group, Tonon, 1995

Draft of »Cena«, Tonon, 1995

Prototype »Cena«, Tonon, 1995

On the right
»Parallele« bed on casters, Ligne Roset, 1995

Draft of »Habitus«, Ligne Roset, 1995

»Biblia« dining table, Ligne Roset, 1993, and »Hera« chair, Ligne Roset, 1990

Das Messen mit den Meistern des Bauhauses über die Jahrzehnte hinweg mag ihn bei der Arbeit beflügelt haben. Entstanden ist jedenfalls ein Stuhl, dessen elegante Leichtigkeit an den »Superleggera« von Gio Ponti oder an den kargen Minimalismus von Jasper Morrisons »Plywood Chair« erinnert. Dennoch ist es ein Entwurf von großer Eigenständigkeit. Prototypisch vereinigt der Stuhl, was die besten Entwürfe Malys stets ausgezeichnet hat: die Liebe zur Geometrie, das Streben nach Reduktion, der wirkungsvolle Einsatz verschiedener Materialien in sorgfältiger Verarbeitung. Mit der Produktion eines neuen, schmutzabweisenden Baumwollgewebes für das Geflecht erfüllt das Sitzmöbel zudem einen innovativen Anspruch und bietet ungeahnten Komfort.

Sechsunddreißig Monate hat die Entwicklung gedauert, um viele Nuancen und Details wurde gerungen. Letztlich jedoch ist er das Ergebnis von dreißig Schaffensjahren und zugleich ganz und gar ein Produkt seiner Zeit, eine Ikone der neunziger Jahre. Das macht seine Qualität, das macht seinen Erfolg aus.

Der Weg der Einfachheit ist für Peter Maly noch lange nicht zu Ende. Immer noch ist er auf der Suche. Immer noch hat er das Gefühl, Neues finden zu können in dem Spiel mit den archetypischen Formen Quadrat, Dreieck und Kreis. Dies ist vielleicht eines der überraschendsten Ergebnisse der »Neuen Einfachheit«: daß sie mit jedem gelungenen Entwurf demonstriert, wieviel ungeahnte Spielarten der vermeintlich so leicht überschaubare Bereich des »Einfachen« immer noch hervorbringt – nachdem ihn so mancher schon als »ausgereizt« hatte verbannen wollen. »Die Möglichkeiten«, da ist Peter Maly sehr zuversichtlich, »sind beileibe noch nicht ausgeschöpft.«

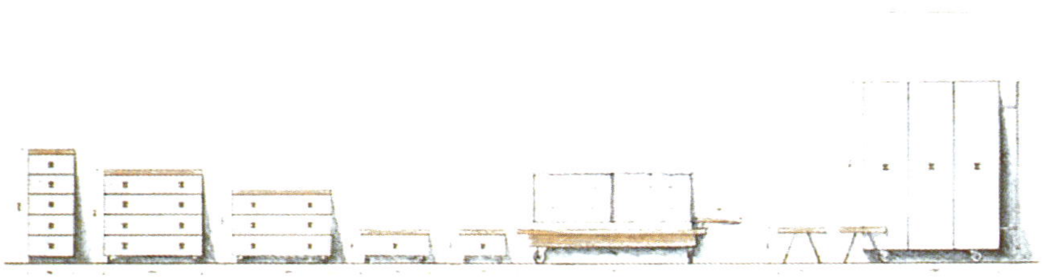

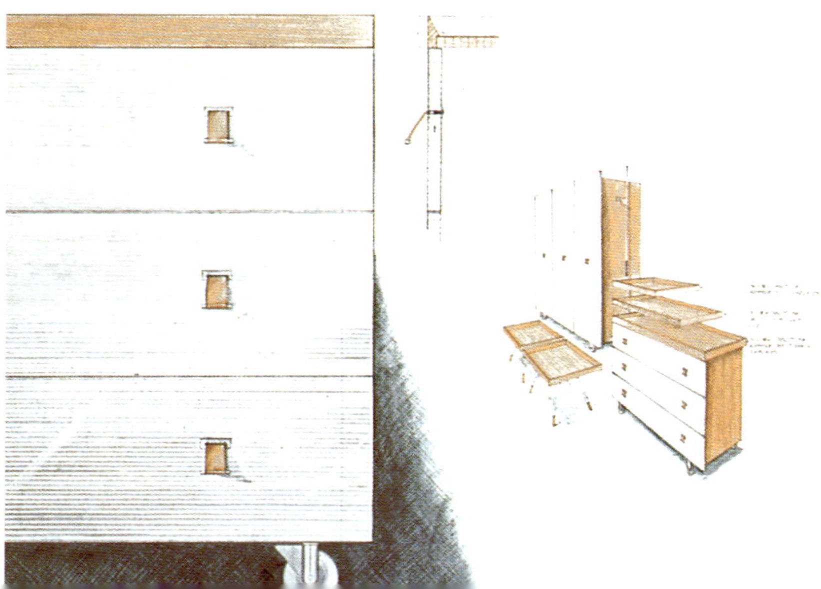

Linke Seite
Sofa »Incognito«, Ligne Roset, 1995

Hochlehnsessel »Habitus«, Ligne Roset, 1995

Entwurf zu »Parallele«, Ligne Roset, 1995

Rechte Seite
Mobile Kommoden und Schränke »Parallele«, Ligne Roset, 1995

Couchtisch »Presto«, Ligne Roset, 1995

Stuhl »Hera«, Ligne Roset, 1990

On the left
»Incognito« sofa, Ligne Roset, 1995

»Habitus« high-backed armchair, Ligne Roset, 1995

Draft of »Parallele«, Ligne Roset, 1995

On the right
»Parallele« chests of drawers and cupboards on casters, Ligne Roset, 1995

»Presto« coffee table, Ligne Roset, 1995

»Hera« chair, Ligne Roset, 1990

Helmut Lübke

Visionen vom besseren Leben – Träume vom schöneren Wohnen

Verantwortliches Handeln und ganzheitliches Denken sind gefragt. Allen ist heute bewußt, daß wir nicht durch kurzfristige Erfolge glücklich werden können. Peter Maly denkt stets über den Tag hinaus. Und das macht seine Arbeit für ein Unternehmen wie COR so wertvoll.

Den großen Zusammenhängen galt schon seine Aufmerksamkeit, als wir uns vor über 25 Jahren kennenlernten. Sein Umweltbewußtsein war damals bereits ebenso ausgeprägt wie seine Verantwortung gegenüber denjenigen, die sich mit seinen Möbeln einrichten. Das gesamte Schaffen des Designers ist von intellektueller Gesinnung und vernetztem Denken geprägt.

Wenn es um die Marktfähigkeit seiner Produkte geht, stellt Peter Maly hohe Ansprüche. Er übernimmt gegenüber dem Unternehmen besondere Verantwortung: Die Arbeitsplatzsicherung ist ihm ein ebenso ernsthaftes Anliegen wie der langfristige wirtschaftliche Erfolg des Unternehmens.

Visions of a better life – dreams of more beautiful interiors

In today's world, circumstances call for responsible action and holistic thinking. Everyone is aware of the fact that we cannot find happiness through short-lived success. Peter Maly always thinks one day ahead, which is why his work is so valuable for a company like COR.

He had already begun think in these larger contexts when we met over 25 years ago. At that time, his concern for the environment was just as pronounced as his sense of responsibility to those who decorated their homes with his furniture. An intellectual attitude and an interlinked approach characterize the entire oeuvre of this designer. Peter Maly is highly demanding when it comes to the marketability of his products. He takes on special responsibility with respect to his client: for him, ensuring jobs is just as important as the company's long-term economic success.

We do not consider our association with Peter Maly to be a typical designer-customer relationship. Rather, we have always explored together whether this or that way of thinking is the right way to handle a new theme. After we have come to a mutual agreement, he then turns to his creative work, highly conscientious in all he does. There are many dimensions to good design, which nonetheless yield an integrated whole. Irrespective of whether the goal for a new design was determined in collaboration with the customer, or whether Peter Maly formulates a goal based on his own observation of the Zeitgeist, the holistic aspect always forms the basis for his thorough approach.

COR-Showroom in Ludwigsburg, COR, 1994

COR showroom in Ludwigsburg, COR, 1994

Peter Maly ist für uns kein »Auftragnehmer« im üblichen Sinn. Niemals haben wir ihm einen konkreten Auftrag erteilt. Vielmehr haben wir stets miteinander ausgelotet, ob bei der Bearbeitung eines neuen Themas dieser oder jener Denkansatz richtig ist. Einvernehmlich mit uns ging er anschließend mit großer Gewissenhaftigkeit an seine kreative Arbeit.

Gutes Design hat viele Dimensionen. Und doch fügt es sich zu einer Einheit. Ob die Zielrichtung für ein neues Design gemeinsam erarbeitet worden ist, ob Peter Maly aus eigener Beobachtung des Zeitgeistes heraus ein Ziel formuliert, immer ist der ganzheitliche Aspekt die Basis seiner gewissenhaften Arbeit.

Nicht für jeden Designer ist es selbstverständlich, daß ein Sessel auch bequem sein sollte. Zu oft wird der Sitzkomfort der schönen Form geopfert. Nicht so bei Peter Maly. Er sucht stets Aspekten der Ergonomie Rechnung zu tragen. Sein Ziel: das entspannte und unverkrampfte Sitzen. Es steht während des gesamten Entwicklungsprozesses im Vordergrund.

Nicht das Unikat interessiert ihn. Seine Aufmerksamkeit gehört dem Industrieprodukt. Das ist Maxime seiner Designarbeit; ihr trägt er schon im Frühstadium eines jeden Entwurfs Rechnung. Dabei geht er keineswegs ausgetretene Pfade; übliche Fertigungsmethoden stellt er gelegentlich in Frage. Er fordert von uns Herstellern innovative Verfahren – stets aufs neue.

The fact that a chair should be comfortable as well as attractive is not always self-evident in the mind of a designer. All too often, comfort in sitting is sacrificed to appealing form. This is not the case with Peter Maly. He always endeavours to give due consideration to ergonomic aspects. His goal: relaxed, comfortable sitting, a goal which is always in the foreground during the entire development process.
He is not interested in creating originals, but rather devotes his attention to the industrial product. This is the maxim of his design work, which he keeps in mind from the moment he begins each design. However, he does not merely keep to the beaten track; at times he calls traditional production methods into question. He demands that we, the producers, constantly come up with innovative processes – repeatedly.
As he sees it, it goes without saying that ecological aspects are to be given careful consideration when creating a product. Maly knows that long-term designs and lasting quality are essential prerequisites for ecologically sound furniture. Maly would never dream of creating throwaway products, for that would be extremely irresponsible. In this respect, we can only agree with him.
On this basis, we work together toward perfecting a piece of furniture, always careful not to lose sight of the vision, to give the concept the consideration it deserves. And I presume – perhaps a bit brashly – that Maly's constructive co-operation with us is

COR-Showroom in Ludwigsburg, COR, 1994

Isometrische Darstellung des COR-Showrooms in Ludwigsburg, COR, 1994

COR showroom in Ludwigsburg, COR, 1994

Isometric outline of the COR showroom in Ludwigsburg, COR, 1994

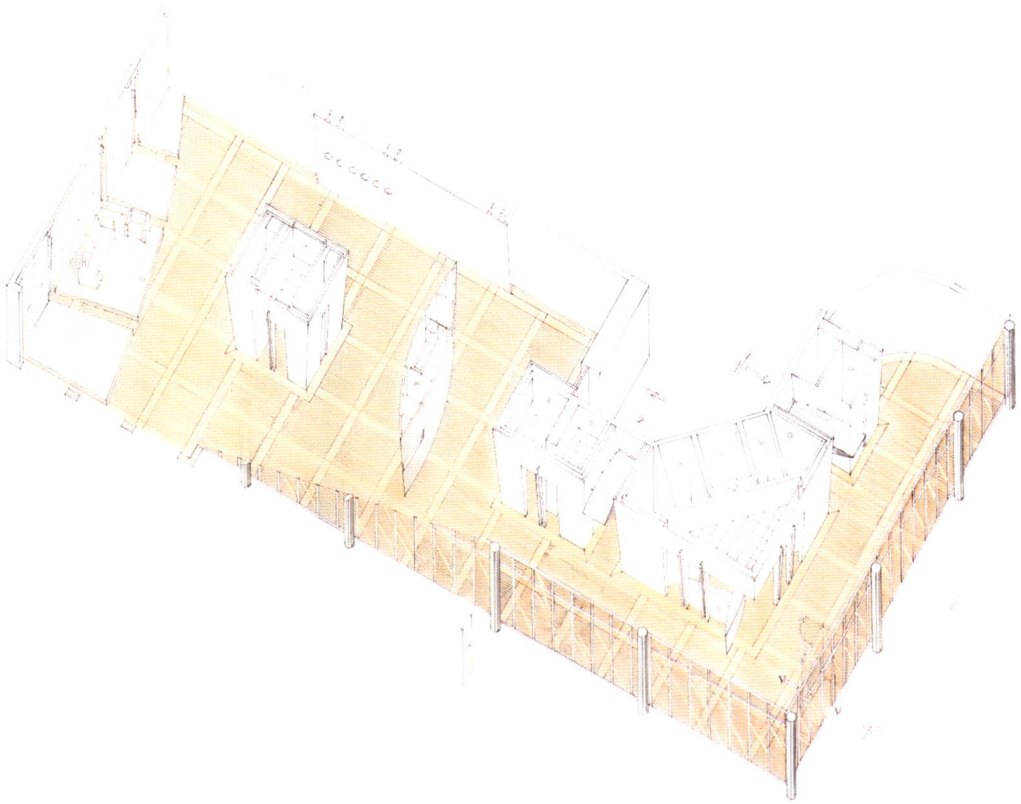

Für ihn ist es selbstverständlich, daß ökologische Aspekte in der Fertigung des Produkts eine gewissenhafte Beachtung finden. Maly weiß: Langzeitdesign und Langzeitqualität sind wesentliche Voraussetzungen für ein ökologisch wertvolles Möbel. Wegwerfprodukte herzustellen, käme Maly nicht in den Sinn, denn das wäre in höchstem Maße verantwortungslos. Und da können wir ihm nur zustimmen.
Vor diesem Hintergrund arbeiten wir gemeinsam an der Vervollkommnung eines Möbels, stets bedacht, die Vision nicht aus den Augen zu verlieren, der Idee Rechnung zu tragen. Und ich mutmaße – vielleicht ein wenig selbstsicher –, daß ihm die konstruktive Gemeinsamkeit mit uns Freude zu machen scheint und seine Kreativität steigert. Das überträgt sich auf alle: Malys Teamfähigkeit beflügelt die Techniker zu neuen Lösungen.
Auch soll nicht unterschätzt werden, daß bei aller konstruktiver Kooperation die formale Anmutung eines Möbels eine große Rolle spielt. Wir als Hersteller wissen, daß der gelungene Auftritt den ersten Impuls bei der Kaufentscheidung der Kunden auslöst.

COR-Messestand zur Internationalen Möbelmesse in Köln, COR, 1995

Isometrische Darstellung des COR-Messestandes in Köln, COR, 1995

COR trade fair stand at the International Furniture Trade Fair in Cologne, COR, 1995

Isometric outline of the COR trade fair stand in Cologne, COR, 1995

a pleasure for him, and enhances his creativity. This is a feeling which affects everyone involved: Maly's penchant for teamwork inspires the technicians to find new solutions. The formal appeal of a piece of furniture plays a role which cannot be underestimated in any constructive joint effort between designer and producer. As producers, we know that customers are more likely to purchase products that have been presented in an attractive way.

This is of great importance, for the market is capricious and does not always instantly reward designers and producers for their efforts. For example, only now are a number of Bauhaus classics gracing every superior interior design – 60 years after they were produced. From time to time, one must be able to wait for success. This is not always easy, as our partners, in particular dealers, are under pressure to make a profit: a new product in the furniture store takes up showroom space and must pull in sales. On the other hand, if the enthusiastic speciality dealer has faith in the designer and the producer, and if everyone involved has enough staying power to persist until the product has attained a certain degree of familiarity among consumers, a Peter Maly design can become a classic.

On the one hand, the market always demands something new. On the other, if a novel form is too unfamiliar, there is a chance it will not be accepted. Thus, it is necessary to find the happy medium between these two poles, which means asking the following question before beginning work on a design: What justifies our adding another armchair to the thousand models that are already on the market? To which the answer might be: a new series must contain a true innovation in some area.

Like other companies, COR is only capable of maintaining a disciplined approach when marketing a new product if we cultivate the image of our models. Thus, for example, the »Combra« chair and dining-table series, designed in 1976 by Peter Maly, met with success in our fast-lane age due to ongoing, subtle improvements and supplementations. Based on his own knowledge, and inspired by customers' ideas, Peter Maly took the original model of his dining room chair and – through additional variations to the armrests – created an entire chair series. The rectangular table was soon complemented by a round pull-out table; a corner bench, cabinets and shelves later rounded out the system. »Combra« remained contemporary thanks to materials, wood types and surfaces that were constantly updated.

Our business partners truly appreciate this approach to cultivating models and keeping a series up to date: the creativity, know-how, and continuity of a design ensure the longevity of our partnership.

All the same, Peter Maly does not always bask in the limelight. Behind the scenes of the interior design culture as well – in obscurity – he gives remarkable fresh impetus and influences innovative interior design, to a certain extent as a co-ordinator of events. The Hamburg designer considers it of utmost importance to impart his ideas convincingly to others, from the producer to the interior design consultant, to the

Die Qualitätsmöbel von heute sind die Klassiker von morgen. Journalisten schreiben gern solche Sätze, wenn es um Interieur und Einrichtung geht. Sinn ergibt das im Zusammenhang mit Peter Maly und seinem Sessel »Zyklus« für COR. Mit diesem Sitzmöbel hat sich Maly für die »Hall of Fame« der Designer qualifiziert.
Und typisch für Denken und Vorgehen dieses Mannes ist »Zyklus« außerdem: ein Möbel, reduziert auf die Kraft geometrischer Urformen, das von einer ganz eigenen, dynamischen Art des Sitzens erzählt. Die Arbeit Peter Malys auf »Zyklus« zu reduzieren, wird gerade ihm nicht gefallen und trifft doch die Wahrheit.

Today's quality pieces of furniture are tomorrow's classics. Journalists like to write aphorisms like this when it comes to interior design and home furnishings. Nevertheless, it makes sense in connection with Peter Maly and his »Zyklus« chair for COR. Maly has qualified himself for the designer »Hall of Fame« with »Zyklus«. Furthermore, »Zyklus« is also typical of this man's mode of thought and action: reduced to the strength of basic geometric forms, it conveys a very unique, dynamic way of sitting. Peter Maly of all people will not want his work to be considered solely in terms of »Zyklus«, and yet it is the truth.

Dirk Meyhöfer

Journalist und Architekturkritiker in Hamburg

Journalist and architectural critic in Hamburg, Germany

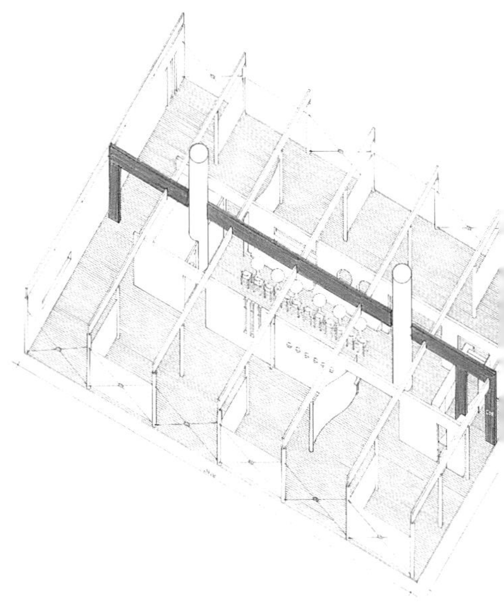

Das ist wichtig. Denn der Markt ist launisch und belohnt nicht immer sofort, was sich Designer und Hersteller vorgenommen haben. Einige Bauhaus-Klassiker beispielsweise zieren erst heute – sechzig Jahre nach dem Entstehen – jede bessere Einrichtung. Man muß also von Zeit zu Zeit auf Erfolg warten können. Das ist nicht immer leicht, da auch unsere Partner, vor allem der Handel, unter Erfolgsdruck stehen: Ein neues Produkt im Einrichtungshaus bindet Verkaufsfläche und muß Umsatz bringen. Hat der begeisterte Fachhändler hingegen Vertrauen zum Designer und Hersteller und verfügen alle Beteiligten über das nötige Durchhaltevermögen, bis ein gewisser Grad an Vertrautheit erreicht ist, dann kann ein »Peter Maly« zum Klassiker werden.
Einerseits fordert der Markt permanent Neues. Andererseits: Ist das Neue in der formalen Anmutung nicht vertraut, wird es mitunter kaum akzeptiert. Es gilt also, das rechte Maß zwischen diesen beiden Polen zu finden und sich immer wieder vor jeder Entwicklungsarbeit die Frage zu stellen: Was legitimiert uns, den Tausenden Sesselmodellen auf dem Markt noch ein weiteres hinzuzufügen? Eine mögliche Antwort lautet: Ein neues Programm muß auf irgendeinem Gebiet eine echte Innovation beinhalten.
Wie andere Unternehmen ist auch COR nur dann in der Lage, beim Vermarkten neuer Produkte Disziplin zu bewahren, wenn wir Modellpflege betreiben. So hat beispielsweise das Stuhl- und Eßtischprogramm »Combra«, 1976 von Peter Maly entworfen, durch kontinuierliche subtile Verbesserungen und Ergänzungen erfolgreich die schnellebigen Zeiten bestanden. Peter Maly hat aus eigener Erkenntnis und auf Anregungen von Kunden hin aus dem Urmodell seines Eßzimmerstuhls – durch zusätzliche Armlehnenvarianten – ein ganzes Stuhlprogramm geschaffen. Den rechteckigen Tisch ergänzte bald ein runder Ausziehtisch, und später wurde das System durch Eckbank, Schränke und Ablagen komplettiert. Dank ständig aktualisierter Materialien, Holzarten und Oberflächen blieb »Combra« aktuell.

Peter Maly – für mich sehr ideenreich und noch produktiver. Immer wieder überraschend und sehr professionell.

Peter Maly – as I see him, a man bursting with ideas, and more productive than ever. He never fails to surprise us, and is highly professional.

Burkhard Vogtherr

Designer in Mulhouse, Frankreich

Designer in Mulhouse, France

person whose »living« they are to enhance (I am intentionally not using the term »end consumer«, because cultural goods are not consumed, they are, at most, used). Catalogue photographs are an excellent means for doing this kind of convincing. Peter Maly is always successful at adroitly setting the scene for his furniture and creating room installations that do justice to his models. One of his duties as former head interior designer at the magazine »Schöner Wohnen«, was to assist readers when they had questions regarding interior furnishings. The auxiliary materials, room perspectives and building sketches used in those days are the same ones he uses today in designing our catalogue pictures. He is always able to create the appropriate ambience for the various models, thus realizing his overall concepts: concrete utopias, visions of a better life, dreams of higher-quality living. All of these, without exception, are inventions of the imagination. Impressive rooms that are set up only for a limited time, short-lived studio scenery, temporary installations – only the furniture is real and permanent.
Competent photographic design is tremendously stimulating for furniture presentations. It triggers the intended associations and desires, and visualizes superior interior design, thus conveying the original idea behind the furniture.
A presentation at a trade fair can also be convincing. Claus Merbold investigated so-called business-to-business communication and published his findings in a series put out by the »Spiegel« publishing company. This journalist concludes that fairs and exhibitions are of great importance in finding an orientation and seeking specific models, as well as when making contacts with producers and examining the products. People buy at trade fairs. Corporate images gel here as well: experts exchange ideas, discuss their observations, and draw the exhibitor into the discourse.
The fact that COR has often been spoken of favourably at trade fairs can be credited to Peter Maly. After all, he draws on an enormous reserve of ideas for his creations and

Von links nach rechts
Fotoinszenierung für COR (»Sera«), COR, 1994

Fotoinszenierung für COR (»Taurus«), COR, 1994

Fotoinszenierung für COR (»Cocon«), COR, 1994

Fotoinszenierung für COR (»Pando« und »Cirrus«), COR, 1994

From left to right
Photographic presentation for COR (»Sera«), COR, 1994

Photographic presentation for COR (»Taurus«), COR, 1994

Photographic presentation for COR (»Cocon«), COR, 1994

Photographic presentation for COR (»Pando« and »Cirrus«), COR, 1994

Unsere Partner im Handel wissen diese Art der Modellentwicklung und Programmpflege außerordentlich zu schätzen: Kreativität, Kompetenz und Kontinuität des Designs sind Garanten für eine Beständigkeit in unserer Partnerbindung.
Aber nicht immer steht Peter Maly im Rampenlicht. Auch hinter den Kulissen der Wohnkultur – im Verborgenen – gibt er bemerkenswerte Impulse und beeinflußt die innovative Einrichtung – als Regisseur gewissermaßen. Der Hamburger Designer hält es für außerordentlich wichtig, daß seine Ideen auch überzeugend weitervermittelt werden: vom Hersteller über den Einrichtungsfachberater bis hin zum »Wohner« (Ich sage bewußt nicht »Endverbraucher«, denn ein Kulturgut wird nicht verbraucht, es wird allenfalls genutzt.).
Solche Überzeugungsarbeit läßt sich hervorragend mit Katalogfotografie leisten. Peter Maly gelingt es immer wieder, seine Möbel gekonnt in Szene zu setzen und modelladäquate Räume zu inszenieren. Als ehemaliger Chef-Innenarchitekt bei der Zeitschrift »Schöner Wohnen« war er unter anderem dafür zuständig, den Lesern in Einrichtungsfragen zur Seite zu stehen. Damals verwendete Hilfsmittel, Raum-Perspektiven und Bauskizzen, nutzt er heute bei der Gestaltung unserer Katalogbilder. Immer wieder gelingt es ihm, für unterschiedliche Modelle das typgerechte Ambiente zu schaffen und so seine Ideen zu verwirklichen. Es sind konkrete Utopien, Visionen vom besseren Leben, Träume vom schöneren Wohnen. Allesamt Imaginationen – Erfindungen. Eindrucksvolle, aber nur für begrenzte Zeit realisierte Räume, kurzlebige Atelierkulissen, temporäre Installationen – allein die Möbel sind echt und werden überdauern.
Für Möbel-Präsentationen ist gekonntes Fotodesign ungemein stimulierend. Es löst die erwünschten Assoziationen und Begehrlichkeiten aus, visualisiert beste Innenarchitektur und vermittelt auf diese Weise die ursprüngliche Idee.
Auch eine gute Messepräsentation kann überzeugen. Claus Merbold hat die Business-to-Business-Kommunikation erforscht und seine Ergebnisse in einer »Spiegel«-Verlagsreihe veröffentlicht. Der Publizist kommt zu dem Ergebnis, daß Messen und Ausstellungen bei der Orientierung, bei der Suche sowie in der Kontakt- und Prüfphase

Linke Seite, unten
Modell des Ligne Roset-Messestandes zur Internationalen Möbelmesse in Köln, Ligne Roset, 1995

Linke Seite, oben und rechte Seite
Ligne Roset-Messestand in Köln, Ligne Roset, 1995

Below left
Model of the Ligne Roset trade fair stand for the International Furniture Trade Fair in Cologne, Ligne Roset, 1995

Above left and right
Ligne Roset trade fair stand in Cologne, Ligne Roset, 1995

seldom repeats himself, while simultaneously achieving a formal idiom that is recognisably his own. He is constantly presenting his models in a novel way. Always in a better way. And always with an element of surprise.
The exemplary interior design concepts to be found in Maly's trade fairs stalls have been noted in expert circles for quite some time. Partners in the business come not only to see our new objects, but also to find inspiration in questions prompted by the presentation: how can I improve or update the showrooms in my own furniture store? Dealers experience this for themselves at COR fair stalls: various interior worlds are excellently staged, and the models are displayed at their best.
Our impressive presentations at trade fairs always contribute favourably to our image and our market acceptance. However, the final product is preceded by hard, responsible work. It has never been exactly child's play to hold one's own in international competition, and if insufficient effort has been put into setting the course at such an event, it can trigger serious economic problems for a company.
Peter Maly takes this challenge seriously. After working up the architectural design for the fair stall, he considers the following questions: how can the models be presented optimally? And which variation on this or that product line is our business partner to take for his own display?

Peter Maly – bedeutet für mich zugleich Gisela Maly. Mit beiden verbindet mich seit fast dreißig Jahren das Interesse an der Kunst, überdies Zusammenarbeit und Freundschaft.
Peter Maly zeigt in der ihm eigenen zurückhaltenden Art seine Liebe zu Material und Formenreichtum, während Gisela charmant und phantasievoll die Ergebnisse auf Messen und in Ausstellungen in Szene setzt.
Ihre Offenheit, Neugierde und auch ihre Kritik haben für mich immer Gewicht. Zu ihren präzisen Messevorbereitungen gehört Kunst als I-Tüpfelchen der einfallsreichen und konsequenten Inszenierungen. Diese sind geprägt von einer Arbeitsfreude, die abwägt und verwirft, um schließlich glanzvoll und stimmig zu sein.

For me, Peter Maly also means Gisela Maly. For nearly 30 years, I have shared with both of them an interest in art, as well as a working relationship and friendship.
In his own reserved way, Peter demonstrates his love of material and abundance of forms, while Gisela stages the results in a charming and imaginative way at trade fairs and exhibitions.
Her openness, curiosity and also her criticism are always important to me. In her precise trade fair preparations, art is the crowning touch to her imaginative and consistent installations. These are characterized by a love for her work, which weighs options and discards what is inappropriate, so that the results are exquisite and just right.

Renate Kammer

Galeristin in Hamburg

Gallery owner in Hamburg, Germany

einen immens hohen Stellenwert einnehmen. Auf der Messe wird gekauft. Hier festigt sich aber auch so etwas wie das Image eines Unternehmens: Die Fachbesucher tauschen Gedanken, reden über Beobachtungen und bringen die Aussteller ins Gespräch. Daß COR in der Vergangenheit oft im positiven Sinne Messegespräch war, verdanken wir Peter Maly, schöpft dieser doch aus einem großen Ideenfundus, wiederholt sich selten, hat aber dennoch eine wiedererkennbare Formensprache. Er präsentiert die Modelle immer wieder anders. Immer wieder besser. Stets überraschend.
Die vorbildhafte innenarchitektonische Gestaltung der Maly-Stände ist in Fachkreisen längst bekannt. So kommen Handelspartner nicht nur unserer Neuheiten wegen, sondern auch, um sich in Fragen der Präsentation anregen zu lassen: Wie kann man die Ausstellungsräume im eigenen Einrichtungshaus verbessern oder aktualisieren? Händler erleben es auf den COR-Messeständen: verschiedene Wohnwelten sind hervorragend inszeniert, die Modelle optimal dargestellt.
Unsere eindrucksvollen Messeauftritte bringen uns stets einen Zugewinn an Image und Sympathie. Allerdings geht ihnen eine harte und verantwortungsvolle Arbeit voraus. Im internationalen Wettbewerb mithalten zu können, war noch nie ein Kinderspiel; und für eine solche Veranstaltung die Weichen nicht richtig gestellt zu haben, kann sehr schnell gravierende wirtschaftliche Probleme nach sich ziehen.

»Maison Maly«, Inszenierung im Fotostudio (»Habitus«, »Incognito« und Parallele«), Ligne Roset, 1995

»Maison Maly«, stage set for the photographer's studio (»Habitus«, »Incognito« and Parallele«), Ligne Roset, 1995

These are important questions. After all, we are aware of the fact that people want to buy a series just as it was presented at the trade fair. Trends are monitored closely. Peter Maly always displays great tact when offering advice, for example by thoroughly discussing possible combinations as well as the most divergent materials and colours, integrating them to form a whole, and, finally, declaring it a finished product.
We take a similar approach to our ongoing exhibitions at our group headquarters in Rheda-Wiedenbrück, as well as in Ludwigsburg, Baden-Württemberg. These showroom designs must remain up-to-date over a period of several years, while exuding a workshop feel. Sensitivity and conscientiousness are required in order to develop showroom layouts that fulfil all of these requirements. Just like our trade fair stalls, our permanent exhibitions also offer fresh ideas and inspiration.
As a result of Maly's success with his furniture, the media has shown increasing interest in him as a person. As an observer of public life, the designer is finding more and more opportunities to put forth views in which he takes a critical look at society. He takes advantage of these public appearances to counter widespread clichés.
He never tires of explaining that serious design is not the polished arrangement of surfaces, and that the term »design« has practically become an empty word due to inflated and often incorrect usage. By contrast, he states that good, sincere design – in other words design that is developed with holistic aspects in mind – will serve people for many years in addition to enriching their lives.
A good example of such a developmental process is the story behind the work entitled »Zyklus«. The magazine »md« provided a detailed commentary on the emergence of this work in its November 1983 issue: »Inspired by a piece of doll's furniture! Once upon a time, there was a designer who had a daughter named Laura. She had a doll's house, which of course contained furniture. One day, the father borrowed an armchair from his daughter in order to furnish a small model at a trade fair stand. The producer for whom he had done this was satisfied with the designer's work. And perhaps with himself as well, for another idea occurred to him while he was thoughtfully contemplating the model. He shared this idea with the designer, who thereupon sat down at his drawing board in order to translate this idea into fact: an armchair was to be born, just as light and transparent as Laura's chair, only for big people. To make a long story short, let us go straight to the happy ending, an important element of any story. A long period of development yielded a comfortable armchair and a cozy sofa. Both designer and producer could now sit back in their new, comfortable armchair, satisfied with themselves and the state of the world, for such a »lucky draw« does not happen every day. The furniture is saucy, funny, and at least as imaginative as the jaunty, colourful furniture being produced on the other side of the Alps. And its construction is far more clever, as it is more strongly aligned toward industrial production, meaning that the ›Zyklus‹ furniture can also be afforded by younger people.«

Peter Maly – ein wahrhaft international agierender Mann, ein großer Kenner der industriellen Produktion, fähig, reellen und konkreten Projekten Substanz zu verleihen. Intuitiv erkennt er im voraus die Tendenzen des täglichen Lebens, interpretiert sie als persönlichen und klaren Stil, ohne zu provozieren.
Seine Produkte besitzen die Qualität des technischen Wissens und spiegeln seine Erfahrung wider, derer er sich jedoch nicht bis zum Exzeß bedienen will: ungewöhnliche Tugenden bei einem erfolgreichen Mann wie Peter Maly.
Für mich ist er ein geschickter und konsequenter Designer, dessen Arbeitsergebnisse stimulierend und für die Menschen von großem Interesse sind – besonders für junge Leute, die sich mit Design beschäftigen.

Peter Maly – a truly international professional, a great authority on industrial production, capable of giving substance to real, concrete projects and programs. He adroitly identifies the trends of everyday life before they even occur, interpreting them in a clear, personal style without being overly provocative.
His products are imbued with a high degree of technical knowledge and reflect his own experiences, although he does not want to make excessive use of these – an unusual virtue in a successful man like Peter Maly.
In my opinion, he is a skillful and consistent designer, whose work stimulates and arouses the great interest of others – especially young people who are involved in design.

Paolo Nava

Designer und Architekt in Lissone, Italien

Designer and architect in Lissone, Italy

Peter Maly nimmt diese Herausforderung ernst. Nachdem er den architektonischen Entwurf des Messestandes erarbeitet hat, geht es um die Frage: Wie werden Modelle optimal präsentiert? Und in welcher Variante soll unser Handelspartner dieses oder jenes Programm für seine Ausstellung aufnehmen?
Das sind wichtige Fragen. Denn wir wissen, daß ein Programm gerne so gekauft wird, wie es auf der Messe präsentiert wurde. Trends werden aufmerksam beobachtet. Peter Maly ist immer auch sensibler Berater, beispielsweise wenn es darum geht, Kombinationen, unterschiedlichste Materialien und Farben intensiv zu erörtern, zur Einheit zu fügen und schließlich festzulegen.
Ähnlich gehen wir bei unseren ständigen Ausstellungen im Stammsitz Rheda-Wiedenbrück sowie im baden-württembergischen Ludwigsburg vor. Hier muß eine Showroom-Gestaltung über viele Jahre aktuell bleiben und Workshop-Charakter ausstrahlen. Man benötigt Gespür und Gewissenhaftigkeit, um jene Raumkonzepte zu entwickeln, die all diesen Vorgaben gerecht werden. Auch unsere permanenten Ausstellungen bieten Anregungen und Inspiration – wie unsere Messestände.
Mit dem Erfolg seiner Möbel nimmt das Interesse der Medien an der Person Peter Malys kontinuierlich zu. Immer häufiger kommt der Designer als gesellschafts- und sozialkritischer Beobachter der Öffentlichkeit zu Wort. Bei diesen Auftritten nutzt er die Gelegenheit, häufig verbreiteten Klischees entgegenzutreten.
So wird er nicht müde klarzustellen, daß ernsthaftes Design keine lapidare Oberflächengestaltung und der Begriff »Design« durch inflationäre und häufig falsche Verwendung fast zur Worthülse verkommen ist, daß gutes und ehrliches Design aber, Gestaltung, die unter ganzheitlichen Aspekten entwickelt wurde, dem Menschen lange Zeit dienlich ist und sein Leben bereichert.
Ein gutes Beispiel für einen Entwicklungsprozeß ist die Geschichte des »Zyklus«. Die Zeitschrift »md« kommentierte die Entstehung in ihrem November-Heft 1983 detailliert: »Ein Puppenmöbel stand Pate! Es war einmal ein Designer, der hatte eine Tochter, die Laura hieß. Sie hatte eine Puppenstube, und in dieser Puppenstube gab es natürlich auch Möbel. Eines Tages borgte sich der Vater bei der Tochter einen Sessel, um das kleine Modell eines Messestandes damit zu möblieren. Der Fabrikant, in dessen Auftrag er das getan hatte, war mit der Arbeit seines Designers zufrieden. Viel-

Wenn ich noch irgendeinen Zweifel darüber hatte, wer wohl Hamburgs bekanntester Designer sei, so habe ich ihn endgültig verloren, als ich kürzlich auf der Straße vor Malys Studio erlebte, wie sich zwei junge Radfahrer zuriefen, daß hier Peter Maly wohne.
Diese Bekanntheit erklärt sich nur schwer für eine – bei allem Selbstbewußtsein – so bescheiden auftretende Person wie Peter Maly. Sie ist wohl vor allem ein Ergebnis der guten Öffentlichkeitsarbeit seiner Auftraggeber. Malys Werk hätte ohne diese Popularität ebensosehr Bestand, vertritt es doch jene in Hamburg besonders geschätzte Gestaltungskonzeption, die stets dauerhafter Qualität gegenüber wechselnder Mode den Vorzug gab. »Konvention« taugte hier nie als Vorwurf, und »Avantgarde« um ihrer selbst willen hat nie Ansehen genossen.
Auf Peter Malys Stühlen kann man sitzen.

If I had had any doubts about who Hamburg's most famous designer might be, they were dispelled once and for all upon hearing two young bicyclists in front of Maly's studio call out that Peter Maly lived there.
Self-confidence notwithstanding, this fame is difficult to explain in the case of such a modest person as Peter Maly. It is, above all, the result of the excellent PR work of his clients. Maly's work would have been of lasting importance even without this popularity, as it is representative of an approach to design that is valued in particular in Hamburg, where lasting quality has always been preferred over rapidly changing fashion. Here, »convention« was never a word that could be used as a rebuke, and »avant-garde« for its own sake never enjoyed public esteem.
Peter Maly's chairs make for good sitting.

Hartmut Frank

Professor im Fachbereich Architektur an der Hochschule für Bildende Künste in Hamburg

Professor at the department of architecture, School of Fine Arts in Hamburg, Germany

The background for this story was a presentation of »Zyklus« in Maly's studio, to which the press responded quite favourably. All phases of the developmental process were portrayed using the initial sketches, elaborations executed in perfect detail, and explanations.

The next obvious step was for others to take over this furniture presentation in a similar form. Furniture stores and progressive dealers began to introduce the models in an unusual way: Peter Maly created a sort of portable stage that was easy to assemble expressly for this purpose. »Information, not decoration« was his motto. On both walls of the stage, there were friezes of five meters in width, displaying the complete design story, as well as hand-written texts and design sketches from »Zyklus«. These provided excellent arguments in favour of the design – the achievement was no longer merely purported, it was there in black and white for all to see! And what an opportunity to satisfy the curiosity of design fans. While they were test-sitting »Zyklus«, the design process was presented to them before their very eyes.

Linke Seite
Modell des Mercedes-Benz-Messestandes zur Internationalen Automobilausstellung in Frankfurt am Main, Mercedes-Benz, 1987

Rechte Seite
Entwurf zum Mercedes-Benz-Messestand in Frankfurt am Main, Mercedes-Benz, 1987

On the left
Model for the Mercedes-Benz trade fair stand for the International Automobile Exhibition in Frankfurt am Main, Mercedes-Benz, 1987

On the right
Draft of the Mercedes-Benz trade fair stand in Frankurt am Main, Mercedes-Benz, 1987

Mercedes-Benz-Messestand zur Internationalen Automobilausstellung in Frankfurt am Main, Mercedes-Benz, 1987

Mercedes-Benz trade fair stand for the International Automobile Exhibition in Frankfurt am Main, Mercedes-Benz, 1987

leicht auch mit sich selbst; denn er hatte, als er das kleine Modell sinnend betrachtete, auch noch eine Idee. Er teilte sie dem Designer mit ..., und dieser setzte sich an seinen Zeichentisch, um die Idee in die Tat umzusetzen: Ein Sessel sollte es werden, genau so leicht und transparent wie Lauras Sessel, nur eben für die Großen. Um das Ganze ein bißchen abzukürzen, kommen wir gleich zum Happy End, wie es sich für jede Geschichte gehört. Nach langer Entwicklungszeit waren aus der Idee ein bequemer Sessel und ein gemütliches Sofa geworden. Beide, der Designer und der Fabrikant, konnten sich nun in ihrem neuen, komfortablen Sessel zurücklehnen und mit sich und der Welt zufrieden sein; denn so ein ›guter Wurf‹ gelingt nicht jeden Tag. Das Möbel ist frech, lustig und mindestens genauso phantasievoll geworden, wie die ›flippig‹ bunten Wohnobjekte jenseits der Alpen. Und seine Konstruktion ist weitaus pfiffiger; denn sie ist stärker auf industrielle Produktion abgestimmt, so daß ›Zyklus‹, so heißt das Möbel, auch für jüngere Leute erschwinglich ist.«

Soweit die »md«. Der Hintergrund dieser Geschichte war eine Präsentation des »Zyklus« im Malyschen Atelier, welche von der Presse sehr positiv aufgegriffen wurde. Anhand der ersten Skizzen, der perfekten Detail-Ausarbeitungen und Erläuterungen wurde der Entwicklungsprozeß in allen Phasen dargestellt.

Es lag nahe, diese Möbelpräsentation in ähnlicher Form zu übernehmen. Einrichtungshäuser und progressive Handelspartner demonstrierten eine ungewöhnliche Modelleinführung.

Peter Maly schuf eigens zu diesem Zweck eine transportable, leicht aufstellbare Koje. Information statt Dekoration war seine Devise: Auf den beiden Kojenwänden fand sich ein fünf Meter breiter Plakatfries mit der vollständigen Design-Story, handschriftlichen Texten und Entwurfszeichnungen von »Zyklus«. Welche Argumentationshilfe, anhand derer die Design-Leistung gar nicht erst behauptet werden mußte, sondern schwarz auf weiß bewiesen werden konnte! Und was für eine Gelegenheit, die Neugierde von Design-Fans zu befriedigen. Während einer Sitzprobe im »Zyklus« wurde ihnen der Entwurfsprozeß vor Augen geführt.

Of course, such an event cannot take the place of our highly successful showing at the furniture trade fair in Cologne in 1984. There as well, it was Maly who designed our stall in the form of an atrium – and placed »Zyklus« right in the center!
And that was by no means the last appearance of the unique armchair. It was admired on television, had a cameo role in the cinema film »Männer«, adorned title pages of magazines, and was to be found in advertisements as well as numerous documentaries on interior design. This was furniture with a career. »Zyklus« contributed greatly to our image, and today is one of our most famous models; it is without a doubt our most striking.
Despite the excellent resonance in the media, dealers were at first unwilling to take »Zyklus« seriously. We often faced rejection, even derision. Thus, the initial business results were quite modest. However, we refused to give up. We stood by our model and later reaped the rewards: in 1984, »Schöner Wohnen« elected »Zyklus« »Furniture of the Year«, and included the product in the book »Modern Classics« – rightly so.
There is an expression in the north of Germany: »One swallow does not make a summer«. Translated, this might read: one good model does not make for a design-oriented company. One can only speak of a discerning company culture if the various ways of thinking and convictions in a business are in harmony and essentially congruent with one another. If, in co-operating with creative free-lance workers and consultants, the company promotes a team spirit which is also supported by mutual trust,

Im Zuge meiner Arbeit begegne ich allen wichtigen Möbel- und Industrie-Designern, zu denen ich Peter Maly auf jeden Fall zähle. Stets tritt seine ungewöhnliche gestalterische Fähigkeit in den Vordergrund und nicht seine Person. Das findet man bei guten Designern heutzutage selten. Peter Maly entwirft für ein Unternehmen nicht nur hervorragende Produkte, sondern beschäftigt sich auch intensiv mit der Firmenphilosophie, mit Fragen der corporate identity und mission statement. Das Unternehmen profitiert von der gemeinsamen Sache, dem »guten Design«. Der Inhaber mag das »Herz« sein, Peter Maly aber ist die »Seele«. Beides ist notwendig, um lebendig zu sein.

In my work, I meet all the most important furniture and industrial designers in the world, and Peter Maly is certainly one of them. His design prowess far exceeds his ego, and that is a rare quality in most good designers today. When Peter Maly works together with a firm, he not only designs great products for them, but moulds the thought process for the entire company, from corporate identity to the mission statement. The company thrives because there is a common cause – »good design«. The company owner may be the »heart«, but Maly is the »soul«, and you need both to be alive.

Nasir Kassamali

Inhaber und Geschäftsführer von Luminaire in Miami, USA

Owner and President of Luminaire in Miami, USA

Eine solche Aktion kann natürlich nicht den großen Auftritt während der Möbelmesse in Köln 1984 ersetzen. Peter Maly gestaltete auch unseren Messestand. Er nutzte die Form eines Atriums – mit »Zyklus« im Mittelpunkt!
Und das blieb keineswegs der einzige Auftritt des neuartigen Sessels: So war »Zyklus« im Fernsehen zu bestaunen, spielte im Kinofilm »Männer« mit, schmückte Titelseiten von Zeitschriften, war in Werbung sowie zahlreichen Wohn-Reportagen präsent. Ein Möbel macht Karriere. »Zyklus« hat sehr zu unserem Image beigetragen und ist heute eines unserer bekanntesten Modelle, das markanteste allemal.
Trotz der großen Resonanz in den Medien wurde »Zyklus« aber zunächst nicht von allen Händlern ernst genommen. Wir sahen uns häufig mit Ablehnung, gar mit Spott konfrontiert. So war der wirtschaftliche Erfolg anfangs sehr bescheiden. Doch wir gaben nicht auf, hielten am Modell fest und wurden belohnt: »Schöner Wohnen« kürte »Zyklus« 1984 zum Möbel des Jahres und nahm das Produkt in das Buch »Moderne Klassiker« auf – zu Recht.
Im Norden sagt man: »Eine Schwalbe macht noch keinen Sommer«; ich übersetze: Ein gutes Modell allein macht noch kein designorientiertes Unternehmen. Erst wenn alle Gesinnungswerte eines Betriebs übereinstimmen und im Wesen deckungsgleich sind, kann man von einer anspruchsvollen Firmenkultur sprechen. Verkörpert das Unternehmens-Team zusammen mit den freien kreativen Mitarbeitern und Beratern den einheitlichen Geist, der zudem von gegenseitigem Vertrauen getragen ist, wird auch der wirtschaftliche Erfolg nicht ausbleiben. So gesehen sind Peter Malys Produkt- und Fotodesign, Messestand-Entwürfe und Showroom-Gestaltungen auch wesentliche Bausteine unserer Firmenidentität.
Gerade in einem von irrlichterndem Zeitgeschmack oft irritierten Geschäft wie dem unseren bedarf es steter Geradlinigkeit und Orientierung. Kontinuität in der stilistischen Aussage und durchgängige Einheitlichkeit in den verschiedenen Präsentationsformen sind wichtige Voraussetzungen für ein prägnantes Unternehmensbild.
Ich schätze Peter Malys Wesen und seine Persönlichkeit: Oberflächliche Unverbindlichkeiten lehnt er ab – sowohl in der zwischenmenschlichen Kommunikation als auch in professionellen Fragen. Er äußert sich nur, wenn er Wesentliches beizutragen hat. Wir finden zwar in der Regel schnell ein gemeinsames Ziel, aber Peter Maly geht selten den einfachen Weg. Nie macht er es sich leicht. Er arbeitet ausdauernd – bis an die Grenzen der Belastbarkeit, ist dabei äußerst selbstkritisch. Weder vorschneller Applaus noch zeitlicher Druck können ihn beirren auf seinem Weg zur Perfektion. Seine selbst gesetzten Maßstäbe sind hoch. Erst wenn ihm die eigene Arbeit vollkommen erscheint, ist er bereit zu sagen: »Ja, so muß es gehen!«

Ich kannte Peter Maly bereits als bekannten und international erfolgreichen Designer und hatte über viele Jahre seine Ideen und Entwürfe, die er mit den besten Herstellern der Möbelindustrie verwirklicht hatte, verfolgt. Schon lange war es mein Wunsch, mit diesem Designer in Kontakt zu kommen, der mit seiner Kreativität meiner Vorstellung von modernem Möbeldesign entsprach.
So nahm ich mir ein Herz, kontaktierte und bat ihn, für meine Firma ein Stuhl- und Tischkonzept zu entwickeln.
Peter Maly hat nicht lange gezögert, wir konnten bereits zwei Wochen später einen ersten Termin in seinem Hamburger Studio vereinbaren. Der erste Kontakt war beiderseits etwas schüchtern – man mußte sich erst kennenlernen. Beim zweiten Treffen präsentierte er mir bereits die Entwürfe für »Cena«. Ich wußte sofort, daß mein Gefühl mich nicht im Stich gelassen hatte: Dieser Designer besitzt ein außergewöhnliches »feeling«!
Wir fertigten die ersten Prototypen an, und diesmal kam er zu uns nach Italien. Nach einem anstrengenden Arbeitstag fuhren wir zu einem Restaurant in die Hügel von Friaul. Bei Rotwein und gutem Essen kamen wir uns auch persönlich schnell näher. Wir diskutierten über Design und Möbel, aber auch über persönliche Dinge und merkten überhaupt nicht, daß wir – bereits nach Mitternacht – die letzten Gäste waren.
Peter Maly hat mir an diesem Abend das Du angeboten, und ich habe mich darüber sehr gefreut. Denn für mich entsteht der Erfolg einer Zusammenarbeit nicht nur durch geschäftliche Beziehungen – es muß auch Spaß machen. Ich glaube, ich kann auch im Namen von Peter sagen: Uns beiden macht es Spaß!

I was already familiar with Peter Maly as a famous, internationally successful designer, and for many years I had followed the ideas and designs he realized in collaboration with the best furniture producers in the industry. I had wanted to establish contact with this designer for quite some time, since the creativity reflected in his work corresponded to my own ideas on modern furniture design.
So I took heart and contacted him, asking him to develop an integrated chair and table concept for my company.
Without much hesitation, Peter Maly agreed – we arranged a first appointment for just two weeks later at his studio in Hamburg. We were both a bit shy at this first meeting – we had to get to know each other first. At our second meeting, he was able to hand me the designs for »Cena«. I knew immediately that my instincts had not betrayed me: this designer possesses an extraordinary feeling for his work!

»Wenn alles getan ist...«, Helmut Lübke (COR) und sein Designer

»When all is said and done...«, Helmut Lübke (COR) and his designer

there is a very good chance of achieving economic success. Seen in this light, Peter Maly's designs for products, photos, fair stalls, and showrooms are an essential component of our corporate identity.

Particularly in a line of business like ours, which must often come to terms with an extremely fickle prevailing taste, it is necessary to pursue a clear course and find a suitable orientation. A consistent stylistic statement and uniform presentations are important prerequisites for an effective corporate image.

I value Peter Maly's being and personality: he rejects superficial, non-committal relationships, both in interpersonal communication as well as in professional concerns. He speaks only when he has something of importance to say.

While we are usually quickly able to determine a common goal, Peter Maly seldom chooses the most straightforward path. He never makes things easy for himself. He works tirelessly, pushing the limits of his endurance, the whole time extremely critical of himself. Neither overhasty applause nor time pressure can distract him on his path toward perfection. The standards he sets for himself are high. Only when his work appears perfect to him can he finally say, »Aha! That's it!«

We produced the first prototypes, and this time he came to us in Italy. After a strenuous day of work, we drove to a restaurant in the hills of Friaul. Over a bottle of red wine and good cuisine, we became better acquainted on a personal level. We discussed design and furniture, but also personal matters, and did not even notice that it was already after midnight and we were the last customers. On this evening, Peter Maly and I began addressing each other with the informal »du«, which pleased me greatly. As far as I am concerned, good business relations are not enough to make for a successful working relationship – it must be enjoyable as well. In this case, I think I can speak for Peter Maly as well: we are both having fun!

Sandro Tonon

Inhaber von Tonon in Manzano, Italien

Owner of Tonon in Manzano, Italy

Petra Schwab

Ein Designer antwortet mit Möbeln

Was ist es, das einen Designer in die Lage versetzt, so kontinuierlich wie Peter Maly erfolgreiche Möbel zu entwerfen? Maly gestaltete immerhin schon 1968 Titelseiten der Zeitschrift »Schöner Wohnen« (siehe S. 17) und entwarf dazu die passenden Op-Art-Möbel. Daß er nach wie vor für viele Firmen Klassiker und zeitgemäße Produkte entwirft, mag vermuten lassen, daß der Mann ein Rezept für seinen Erfolg geheimhält.
Natürlich gibt es ein solches Rezept nicht. Aber wer Peter Maly genau zuhört, erfährt, warum er auch heute das richtige Gespür für Design hat. Neben formalen Designkriterien und Regeln der Geometrie nämlich – für die Malys Arbeit so oft gelobt wurde – widmet sich der Designer praktischen Einrichtungsfragen. Maly beantwortet diese mit neuen Möbelentwürfen. Etwa: »Weil es eine Antwort auf so viel Dinge gibt,« hält Maly das Programm »Menos« (behr) für eines seiner besten Werke der letzten Zeit. Auch das neuste Sofa »Circum« (COR) ist für ihn »eine elegante Antwort auf das Problem, wie man seine Beine hochlegen kann.« Oder er fragt sich (bei einem Auftrag für die Klavierfirma Sauter): »Mit welchem Klavier antworte ich angemessen auf heutige Wohnformen?« und hat große Freude daran, beim Gestalten neuer Showrooms in

A Designer's Answer: Furniture

What is it that enables a designer to design successful pieces of furniture as consistently as Peter Maly has done? After all, Maly designed cover pages for the »Schöner Wohnen« magazine (see p. 17) as early as 1968. He also designed suitable op art furniture for those pages. The fact that to this day Maly has continued to design classic and contemporary products for many firms might lead us to believe that he is secretly keeping a success formula from us.

Of course, there is no such formula. Whoever listens attentively to Peter Maly will hear why today he still has the right intuitions regarding design: apart from formal design criteria and geometric rules, the designer addresses practical problems of interior decorating. Maly's solution is to present new furniture designs. An example: »Because there is an answer to so many things,« Maly considers his furniture line »Menos« (behr) to be one of his best creations in recent years. And he feels his newest sofa »Circum« (COR) is »an elegant answer to the problem of how to put one's feet up.« When working on a commission for the piano manufacturer Sauter, he asked himself: »With which kind of piano can I adequately respond to current forms of living?« Maly also greatly enjoyed designing new showrooms for ligne roset in Moscow and New York and, in doing so, »responded to the historically grown urban architecture by giving a modern answer.«

Maly, the »master of dialogue« (p. 74), reveals himself – especially with his latest designs – as somebody who is a keen and critical observer. He uses both traditional models and development processes of his own designs to pinpoint a design idea. His second guiding idea is to recognize the design development as a continuous process.

Kommode »Menos«. behr, 1996

Chest of drawers, »Menos«. behr, 1996

Moskau und New York (für ligne roset), »sich auf die gewachsene Architektur der Städte einzulassen und eine moderne Antwort zu geben.«
Maly, der »Meister des Dialogs« (S.74), entpuppt sich gerade mit seinen letzten Entwürfen als jemand, der genau beobachtet: Sowohl traditionelle Vorbilder als auch die Entstehungsgeschichten eigener Entwürfe macht er sich zunutze, um Ideen auf den Punkt zu bringen. Sein zweiter Grundsatz: Eine Designentwicklung als fortlaufenden Prozeß erkennen. Das heißt: Eine geniale Idee landet nicht sofort als großer Wurf auf dem Papier, sondern wächst – manchmal sehr lange. Das Schwierigste: Einen Entwurf im richtigen Moment als fertig zu verabschieden. Jedesmal fragt sich Maly: »Zu welchem Zeitpunkt habe ich alles gegeben, um das zu erreichen, was möglich war?« Während dieses Prozesses muß er oft Grenzen überschreiten, denn meist liegt die Realisierung des entworfenen Möbels jenseits des Machbaren. Drumherum tummeln sich viele Vereinfacher, die es, so Maly, »sehr gut mit einem meinen«. Aber ein Designer muß sich mit manchen Ideen einfach durchsetzen können. »Das ist eine seiner wichtigsten Aufgaben«, betont Maly.
Daß Maly neben allen kreativen Zweifeln eine Designentwicklung im Griff hat, zeigt nicht nur sein Erfolgsprogramm »Menos« aus dem Jahr 1996: Die minimalistisch gestaltete Kommode etwa weckt trotz ihrer zeitgemäßen Form die Erinnerung an alte Apothekerschränke und ist das bestverkaufte Teil der Kollektion – obwohl es auch das teuerste ist. Stolz ist Maly auf die maschinengerecht entwickelten Beschläge. Doch gehört es mittlerweile zum Standard bei ihm, für große Kollektionen auch passende Beschläge zu entwickeln.

This means that an ingenious idea is not immediately committed to paper scoring a great hit, but it evolves and grows – sometimes taking a long time. The greatest difficulty is being able to recognize just when to declare a design »complete(d).« Each time Maly asks himself: »At which point did I give everything I have in order to achieve what was possible?«
In the course of this process, he frequently has to stretch limits because in most cases the implementation of that particular furniture design is really beyond what is feasible. He is surrounded by many sympathetic simplifiers who – according to Maly – »mean so very well.« But a designer simply has to be able to assert himself with his idea and push it through. »That is one of his most important tasks,« says Maly.
Despite all his creative doubting, Maly maintains a tight hold on his design developments. And we have – among others – his successful »Menos« line of 1996 to prove it. Despite its minimalist design and contemporary shape, the «Menos« chest of drawers, for example, evokes memories of old apothecaries' cabinets and is the best-selling piece in the collection, although it is also the most expensive. Maly is particularly proud

Von links nach rechts

Das Programm »Menos« ist für Wohn- und Eßzimmer geeignet. Es besteht aus Regalsystemen, Kastenmöbeln, Kommoden und Tisch. behr, 1996

Funktionalität und schlichte Eleganz geben dem »Menos«-Programm Charakter. Die Beschläge wurden speziell für die Kastenmöbel entwickelt. behr, 1996

»Menos« gibt es in den Farben Blutorange, Maisgelb, Mintgrün, Pflaumenblau, Sahneweiß und Kaviar. Seit 1998 auch mit Walnußholz. behr, 1996

From left to right

The »Menos« collection is designed for living and dining rooms. It consists of shelf systems, unit furniture, chest of drawers and table. behr, 1996

Functionality and simple elegance lend the »Menos« line »personality.« The mountings were especially designed for the unit furniture. behr, 1996

»Menos« is available in the following colors: blood orange, maize yellow, mint green, plum blue, cream white and caviar. Since 1998 also with walnut. behr, 1996

Der große Familientisch harmoniert ebenfalls mit Assoziationen: Er orientiert sich am Vorbild alter Küchentische und ihren Schubladen. Das eindeutig moderne Möbel wirkt trotzdem vertraut und hat Erfolg bei Kunden. Maly: »Der große Familientisch ist heute wieder zu einem zentralen Thema geworden.« Nicht zuletzt die Farben der samtigen Lackoberflächen von »Menos« kommen an: Modisch kräftig, aber dennoch so sensibel gewählt, daß sie sich in unterschiedliche Wohnwelten einfügen, sind sie zweifellos Malys langjähriger Erfahrung auf diesem Gebiet zuzuschreiben.
Ganz sanfte Töne schlägt Maly bei der Kollektion »Mato« (COR) an. Natürliche Materialien stehen im Vordergrund: Stein als Tischplatte, natürliche Baumwolle für Bezugsstoff und Gurtbespannung sowie handwerkliche Verarbeitung von Holz und Polsterung. Die Erfahrung aus einem früheren Entwurf setzt Maly 1996 hier gekonnt ein. Die Ästhetik der Gurtbespannung lernte er nämlich beim Thonetstuhl 737 (S.92/93) lieben. Beim neueren Modell »Mato« nun perfektioniert er das Entwurfsdetail: Der dekorative Streifenkontrast von dunklen Gurten und hellem Holz dominiert die ökologisch produzierte Polstergarnitur.
Seiner Vorliebe für die Epoche des Biedermeier folgt Maly bei dem bequemen Eßzimmerstuhl »Rota« (Tonon). Eine klare Kreisform bestimmt das Sitzmöbel, dessen Lehne sich im exakt gleichen Radius wie die Sitzfläche rundet. Respekt vor traditioneller Gestaltung hätte ihn fast davon abgehalten, einen seiner zuletzt sehr geschätzten Aufträge auszuführen: die Neugestaltung von Klavieren (Sauter).

Links
Funktionale Elemente werden dekorativ: Die Gurtbespannung wird zum kontrastreichen Muster auf dem Holzrahmen. COR, 1996.

Unten
Kein Öko-Möbel im rustikalen Sinn, sondern elegant trotz ökologisch korrekter Fertigung: Couch, Sessel und Tisch »Mato«. COR, 1996

Left
Functional details become decorative ones: the webbing makes a pattern of strong contrasts on the wooden frame. COR, 1996

Below
No »ecological« furniture in the rough-and-ready sense of the word, but elegant despite ecologically correct manufacture: sofa, armchair and table »Mato«. COR, 1996

of the production-adapted design of the mountings. To him, it is standard procedure by now to also develop the »coordinate« mountings for large furniture collections. The huge family table, modeled as it is on old kitchen tables with drawers, is also in tune with historic associations. Maly's clearly modern table nevertheless appears familiar and is very popular with private buyers. Maly comments: »The big family table has again become a central theme today.« It is not least the colors of the velvety varnished surfaces of his Menos furniture range that catch on. They are strong, in line with current tastes, yet chosen with enough sensitivity to fit into the most diverse domestic worlds. This is no doubt due to Maly's long years of experience in the field.

For the »Mato« range (COR, 1996) Maly adopted very gentle color values and predominantly used natural materials: stone for table tops, untreated cotton for covers and webbing, skillfully worked wooden frames and upholstery. Here Maly applied all the experience gained from a previous design to good advantage, especially with regard to the aesthetics of webbing, which he learned – and learned to appreciate – from the Thonet chair 737 (pp. 92, 93). For his more recent »Mato« model he perfected the design detail: the decorative contrast of dark girthbands and light wood is the most striking characteristic of this ecologically manufactured living-room suite.

In designing his comfortable dining room chair »Rota« (Tonon), Maly followed his preference for the Biedermeier period (1815–1848). A pure circle is the determining geometric form of this chair, with exactly the same radius for both backrest and seat.

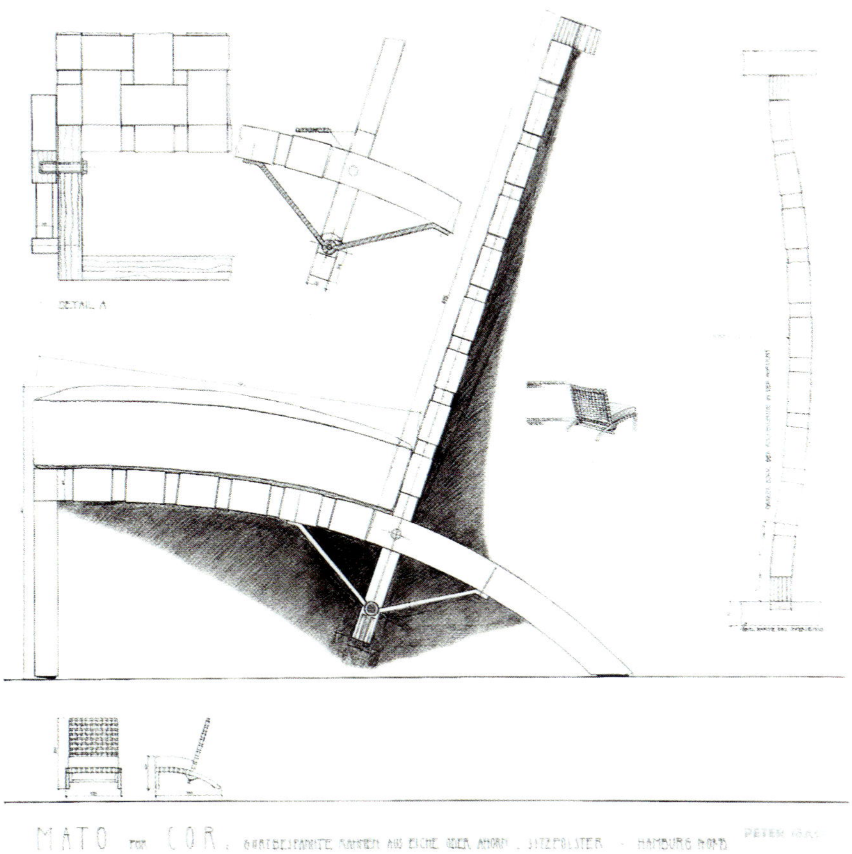

Oben
Die Luxusausführung des Stuhls »Rota« ist aus Vollholz. Dazu gehören aber auch Sperrholz-Varianten und ein Tisch. Tonon, 1996

Unten
Stühle wirken von jeder Seite anders: »Alpha« sieht von vorn aus wie ein robustes »A«, von der Seite wie eine elegante Skulptur. Tonon, 1997

Above
The luxury version of the »Rota« chair is made of solid wood. »Rota« also comprises plywood versions and a table. Tonon, 1996

Below
Chairs look different from every side: from the front, »Alpha« looks like a solid »A«, from the side like an elegant sculpture. Tonon, 1997

Linke Seite oben
Beim Modell »Artes« fügte Maly kleine Art Deko-Elemente ein, die zum Hochglanzlack und vielen Einrichtungen passen. Sauter, 1996

Unten
»Imago« ist die modernste Version eines modernen Klaviers. Frontrahmen aus Birnbaum oder Kirschbaum, Mittelteil Lack. Sauter, 1996

Detail »Vivace«. Sauter, 1997

Rechte Seite
Das Glanzstück unter den Musikmöbeln: Flügel »Vivace« in Ahorn oder schwarzem Hochglanzlack, stets mit Intarsien . Sauter, 1997

Above left
For the »Artes« model Maly added small decorative details that fit in with both the high-polish finish and many different furnishing styles. Sauter, 1996

Below
»Imago« is the ultra-modern version of a modern grand piano. Front: pearwood or cherry, body satin-finished. Sauter, 1996

Detail »Vivace«. Sauter, 1997

Right
The pièce de résistance of Maly's musical furniture: grand piano »Vivace« in maple or black high-gloss varnish, each version with ornamental metal plate band. Sauter, 1997

Feuer und Flamme für Klavier-Entwürfe war Maly erst, nachdem er die Klavierbauer besucht hatte: »Sie beherrschen unglaublich viele Handwerkstechniken und können gut aufs Design eingehen«, schwärmt er heute und gibt mit den ersten drei Klavieren in zeitgemäßem Design wiedermal eine Antwort – auch auf eigene Probleme als Innenarchitekt: »Wenn ich eine Wohnung einrichte, zu der ein Klavier gehört, stellt sich die Frage ›wohin damit?‹ Denn die Formensprache der meisten Klaviere stammt aus den letzten Jahrhunderten.« In solchen Fällen kann Maly heute eines seiner Klaviere empfehlen oder sogar einen modernen Flügel, den er ein Jahr später, 1997, entwickelte.

»Das war der Höhepunkt meiner Arbeit an Musikinstrumenten«, sagt Maly, der das schwere Musikmöbel etwas leichter erscheinen lassen wollte – etwa durch schlanke Metallbeine. Seine Begeisterung für mögliche Techniken wie komplizierte Holzverbindungen, Oberflächenbehandlungen und Intarsienarbeiten verleitete den Puristen sogar dazu, die Handwerker mit einem aufwendigen Dekor herauszufordern: Sie realisierten für ihn ein Streifenband aus Metallplättchen, das die Kurven des Flügelrahmens »Vivace« ziert und das auch bei geschlossenem Deckel sichtbar ist: Erstmals liegt dieser nicht auf dem Rahmen auf, sondern senkt sich in ihn hinein.

Maly's respect for traditional design almost kept him from accepting a commission, the results of which were much appreciated recently: i. e., redesigning pianos (for Sauter). He only became very enthusiastic about piano design after having visited the piano-makers' factory. »They master an incredible number of craft techniques and easily respond to design,« he now enthuses. With his first three modern pianos Maly has also found an answer to his own problems as an interior designer. »When I decorate a flat [house] for people who have a piano, the question is always where to put it. After all, the formal language of most pianos is that of past centuries.« In such cases Maly is now able to recommend one of his own pianos, or even one of the modern grand pianos which he developed a year later, in 1997.

»It was the climax of my work on musical instruments,« said Maly who wanted the heavy pianoforte to appear a little lighter. His enthusiasm for all possible technical details – complex wood joints, refined surface finishes and inlays – even prompted the purist to challenge the craftsmen with an intricate ornament: they had to create a band of small metal plates that follows the curves of the »Vivace« grand piano frame and is even visible with the lid closed because here, for the first time, the lid is not supported by the frame but inserted into it.

Peter Maly und Klaviere: Ich gebe zu, es war nicht meine Idee, und ich wäre vielleicht auch nicht von allein darauf gekommen. Aber dafür habe ich gute Mitarbeiter. Seit 1819 bauen wir Pianos, immer haben wir alles selbst gemacht. Die Klavierbauerfamilie Sauter hat seit fast 180 Jahren die wichtigsten Phasen der Geschichte des Pianofortes miterlebt und mitgestaltet.

Ich bin mir sicher, daß sich meine Vorfahren für durchweg begabte Akteure im Grenzland zwischen Kunst und Handwerk hielten und es wohl auch waren. Wenn ich mir die Exponate in unserem kleinen Museum anschaue, dann fällt mir auf, wieviel Freude am Gestalten hier am Werke war.

Peter Maly wies mit wenigen Handstreichen diesen Teil der Familientradition in die Schranken.

Eine kurze Besprechung mit ihm, dann wenig später die Pläne ... Ungläubiges Staunen, schließlich entspannte Gesichter – das ist ja ganz neu, und: das ist ja toll!

Den Mangel an wirklich neuen Formen haben sich die Klavierhersteller weltweit in seltener Einmütigkeit mit technischen Sachzwängen erklärt, bis Peter Maly alle Einwände entkräftete, ohne die Vorgaben jedoch außer Acht zu lassen.

Peter Maly and piano design. I confess, it wasn't my idea, and I probably would never have thought of it but for my enterprising employees. Our company has been building pianos since 1819, and we always ›did everything ourselves‹. For nearly 180 years now, the Sauter family of piano manufacturers has lived and shaped the most important stages in the history of the pianoforte.

I'm quite certain that my forbears saw themselves as highly talented actors in the borderland between the art of music and the craft of piano-making and they probably were.

When I look at the exhibits in our own small museum, I am struck by the obvious joy of creation, of form-giving, and the love of materials that is evident in these pieces.

With a few strokes of his pencil, Peter Maly did away with that part of our family traditions.

A short consultation – and the submission of 1:1 plans at the second meeting a few weeks later.

First, incredulity, then excitement: well now, this is new – this is brilliant!

With rare unanimity, international piano manufacturers have always attributed the lack of truly new forms to technical requirements, until Peter Maly seemed to just dissolve these constraints, without really evading them in his design.

C. Ulrich Sauter

Mitinhaber der Carl Sauter Pianofortemanufaktur, Spaichingen

Co-owner of Carl Sauter Pianofortemanufaktur, Spaichingen, Germany

Doch nicht nur bei innovativer Gestaltung moderner Flügel, sondern auch bei der Designpflege von laufenden Möbelkollektionen ist Gespür für Material nötig. Das schon in den achtziger Jahren erfolgreiche Programm »duo« (interlübke) etwa ergänzte Maly um matte Glastüren und neue Metallbeschläge, damit es auch in den nächsten Jahren noch Anklang im gehobenen Einrichtungsbereich findet. Solche Re-Designs sind gerade bei weit verbreiteten Kollektionen wichtig, denn die Fabrikation der Grundelemente kann beibehalten werden, obwohl sich ihr Aussehen maßgeblich verändert. Gleiches gilt auch für das Re-Design von »Tecton« (behr), mit dem Maly ein Jahr später, 1997, beauftragt wurde. Hier nutzte Maly die quadratische Aufteilung des Kastenmöbelsystems, um dessen Schlichtheit mittels neuer Gestaltung eine japanische Anmutung zu geben. Geschlossene Schiebeflächen, dezente Beschläge und die richtigen Farben machen es für die heutige Zeit attraktiver.
Eine Kollektion, die mit der Zeit wachsen kann, muß zweifellos schon als langlebiges Produkt angelegt sein. Nur durch Disziplin in der Gestaltung ist es möglich, daß sie »Jahresringe ansetzt«, wie Maly sagt. Er jedenfalls erweiterte seine »Kollektion Peter Maly« (ligne roset) 1997 mit dem leuchtend orangen Sessel »Swing« und dem modular aufgebauten Regal »Top«, sowie den geometrisch gemusterten Teppichen der Serie »Les Carrées« und dem lederbezogenen Stuhl »Graphic« um einen vierfachen Jahresring. Seine gesammelten Werke fanden auf dem von ihm gestalteten Messestand zur Kölner Möbelmesse ihren perfekt abgestimmten Auftritt.
Hier formierten sich der Sessel »Habitus«, die Kommode »Parallele«, das Sofa »Incognito« und der Tisch »Presto« aus den letzten Jahren zu einem harmonischen Ganzen mit »Top«, »Swing«, »Sendai« und »Graphic«. Die vielseitige Einsetzbarkeit der neuen Möbel ist eine von Malys Antworten auf veränderte Wohnbedürfnisse: Flexibilität zählt. Und so ist »Swing« nicht nur ein legerer Sessel mit Fußhocker, sondern kommt gleich in Form eines Eßzimmerstuhls daher, und das Regal »Top« läßt sich nahezu unbegrenzt erweitern. Es setzt sich aus Regalböden und einfachen rechtwinkligen Modulen zusammen, in die sich bei Bedarf Türen integrieren lassen.

Als ich die von Peter Maly im zeitgenössischen Gewand gestalteten Klaviere auf der Frankfurter Musikmesse sah, war mir sofort klar, daß New York City der prädestinierte Ort für deren Präsentation ist. In unseren Ausstellungsräumen in Manhatten können wir nun diese Meisterstücke moderner Gestaltungskunst Architekten, Designern und all denjenigen nahebringen, die sich beim Kauf von den aktuellsten Entwicklungen im Design inspirieren lassen.

When I saw the new Peter Maly line of Contemporary Pianos at the Music Trade Show in Frankfurt, I knew instantly that New York City would be the ultimate location for a Peter Maly Piano Gallery. So, at our Beethoven Pianos NY showroom, I am now presenting these modern art masterpieces to architects, designers, and the audience that seeks out designs on the cutting edge.

Carl Demler

Inhaber von Beethoven Pianos, New York

Owner Beethoven Pianos, New York

It is not only the innovative design of grand pianos, however, but also the consistent revision of available furniture collections that requires a special feeling for and understanding of materials. For example: Maly fitted the »duo« range – already popular in the 1980s – with mat glass doors and new metal mountings so that it will continue to be successful in the up-market furniture business. Such re-designs are important, particularly for widely distributed collections, since they enable the manufacturer to continue the production of basic elements while altering the appearance of the pieces quite considerably. The same applies to the re-designing of »Tecton« (behr) for which Maly was commissioned a year later, in 1997. For it, he used the square grid of the unit furniture system in order to give its simplicity a Japanese flair. Closed sliding planes, reduced mountings and the right colors make the range more attractive to modern buyers.

A collection that can grow over time must certainly be conceived as a long-term product right from the outset. Only the discipline of the design enables it to »develop annual rings,« says Maly. He himself has created four annual rings for his »Collection Peter Maly« (ligne roset) of 1997: i.e., the bright orange »Swing« lounge chair, the geometrically patterned carpets of the series »Les Carrées«, and a leather-covered chair entitled »Graphic«. And he staged a perfect scenario for his collected works with his stand at the 1998 Cologne Furniture Fair.

Linke Seite oben
Materialcollage »duo«-Griffe, Glastüren etc. Interlübke, 1996

Rechte Seite unten
Das Re-Design des Programms »duo« wirkt mit matten Glastüren und neuen Beschlägen wie ein neues Möbel. interlübke, 1991/1996

Above left
Collage of mixed materials, »duo« handles, glass doors, etc., interlübke 1996

Below right
With its mat glass doors and new mountings, the re-design of the »duo« line looks like a new design. interlübke, 1991/1996

Linke Seite, von oben nach unten
Kontrapunkt zur gewöhnlichen Sitzgarnitur: Sessel »Swing« und Hocker gibt es auch als Eßzimmerstuhl und Sofa. ligne roset, 1997

Das intelligente Regalsystem »top« wächst in 34 cm-Schritten mit: Es besteht aus rechtwinkligen Metallteilen, Regalböden und wahlweise Türen. ligne roset, 1997

Rechte Seite, von oben nach unten
Überarbeitung von Sofa »Incognito«, Sessel »Habitus« und Regalsystem »top«. ligne roset, 1997

Entwurf zu »top«. ligne roset, 1997

On the left, from top to bottom
The antithesis of a conventional sofa and armchair group: the »Swing« armchair and stool are also available as a dining-room chair and sofa. ligne roset, 1997

The cleverly designed shelf system »top« grows in 34 cm steps: it consists of rectangular metal parts, shelf boards, and doors on request. ligne roset, 1997

On the right, from top to bottom
Revised version of sofa, »Incognito«, lounge chair »Habitus« and shelf system »top«. ligne roset, 1995/1997

Draft of »top«. ligne roset, 1997

Das gesamte Szenario demonstriert, wie vielseitig eine stilistisch homogene Kollektion sein kann und daß es immer wieder neue Dinge gibt, die ein Bedürfnis befriedigen, das vorher so noch nicht befriedigt wurde. »Ein Designer muß solche Lücken erkennen«, sagt Peter Maly, »er muß die geheimen Wohnwünsche spüren.«
Manchmal erfüllt sich Maly auch selbst Wünsche. Zum Beispiel mit dem Kaminzubehör »conmoto«. So etwas hatte er schon immer gesucht. Doch normalerweise ist sein Motiv nicht, etwas für sich selbst zu machen. Er arbeitet an Wohnideen für Menschen, die er mag, die das Einfache lieben und die sich wohlfühlen wollen. Die etwa den Stuhl »Alpha« (Tonon) zu schätzen wissen, der wie ein solides »A« von vorn aussieht, von der Seite aber eine schwungvolle Linie in der Rückenlehne zu erkennen gibt. Nach unten hin sind die Beine grazil verjüngt und stehen im Kontrast zur Frontansicht. »Wenn ein Stuhl von der Seite so eckig aussieht wie von hinten, ist das langweilig,« stellt Maly fest und weiß einiges zu erzählen über die skulpturhafte Qualität eines Stuhls:
»Im Gegensatz zu einem Schrank muß ein Stuhl von jeder Seite gut aussehen«, erklärt der Designer und betont, daß es deshalb ungeheuer schwierig sei, einen perfekten Stuhl zu entwerfen – abgesehen von den unterschiedlichen Körpern und täglichen Strapazen, die so ein Möbel aushalten muß. Doch funktionale Aspekte versteht Maly stets eng mit ästhetischen zu verbinden. Bei dem Bett »Chess« (Domitex) befestigt er ein textiles Betthaupt einfach an der Wand, statt am Buchen-Bettrahmen. Eine stabile Lösung. Als eine Art Wandbehang dominiert zudem der weiche Stoff – den Maly wie auch die Bettwäsche selbst entwirft – das Bett und läßt Möglichkeiten für neue Inszenierungen zu.

Denk' ich an Peter Maly in der Nacht, bin ich zwar nicht um den Schlaf gebracht, aber meine Gedanken kommen ganz natürlich auf Deutschland. Deutschland, ein gutes Land, in dem sehr viel möglich ist, machbar ist, aber das ich auch als ein Land empfinde, in dem Kreativität weniger geschätzt wird im täglichen Leben als in anderen Ländern. Das auf jeden Fall ist meine Erfahrung. Peter Maly ist sicher einer der produktivsten und engagiertesten deutschen Designer, der sich vorwiegend mit deutschen Produzenten auseinandersetzen muß, die als erstes auf Rentabilität und Profit schauen. Er hat sich gut geschlagen, und sein Einfluß auf den deutschen Geschmack ist allenthalben sicht- und spürbar. Ich wünschte, er würde mehr aufmucken und frecher werden. Er hat die Kraft dazu!

Speaking with a well-known German poet, I would say that when thinking of Peter Maly at night, though I don't lie sleepless, I also end up thinking of Germany.
Germany – a good country where much is possible and feasible – is also a place, I feel, where creativity in everyday life is less appreciated than in other countries. At least that has been my experience. Peter Maly certainly is one of the most productive and dedicated German designers and has to deal mainly with German producers who first and foremost pay attention to rentability and profit. Maly has stood his ground, and his influence on German tastes is visible and palpable everywhere. I wish he would rebel a little more often and show a bit more dash and daring. He has what it takes!

Ingo Maurer

Designer, München

Designer, Munich

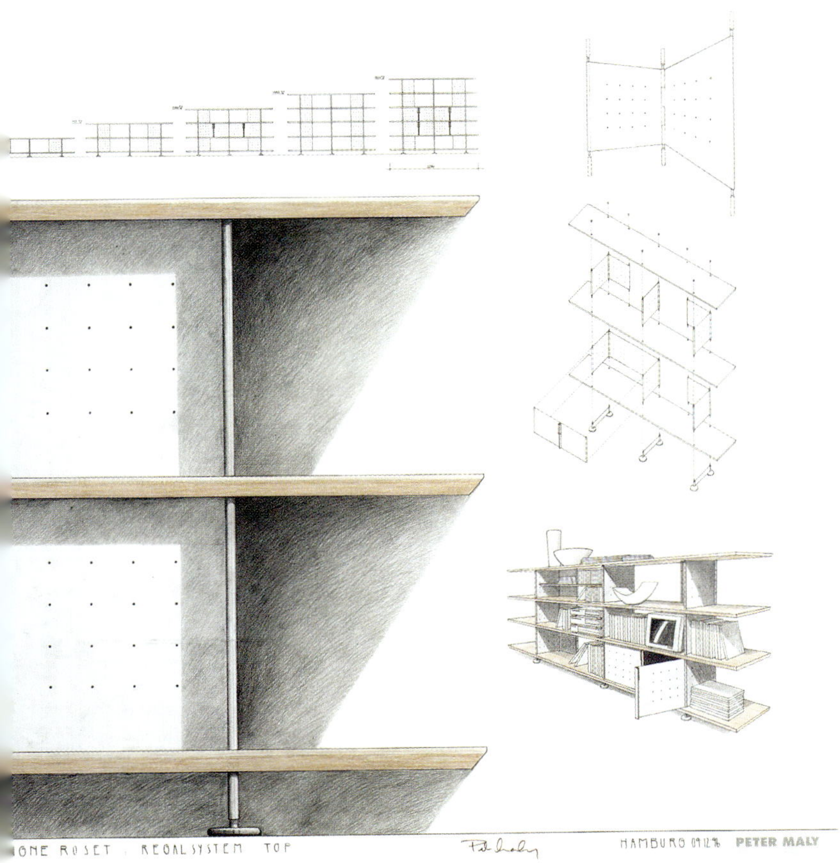

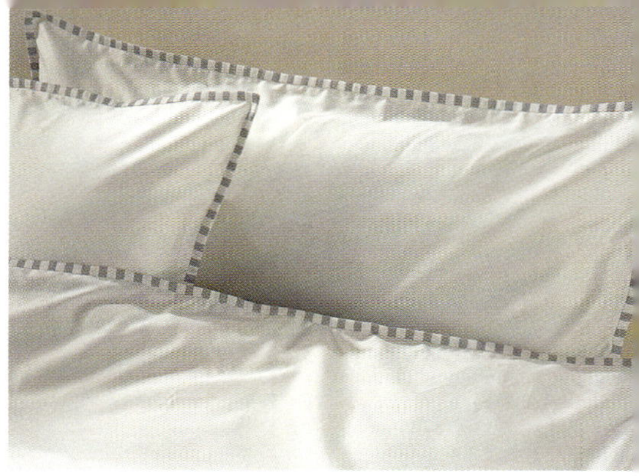

Oben
Zum Bett »Chess« gehört nicht nur das textile, an der Wand befestigte Betthaupt, sondern auch von Maly entworfene Bettwäsche. Domitex, 1997

Links
Leuchtenfamilie aus Mattglas und mattem Stahl: Stehleuchte, Wandleuchte, Tischleuchte in zwei Größen, »Screen«, Anta, 1997

Above
The bed »Chess« is available with fabric bedhead (»wall hanging«) and the bed linen Peter Maly designed especially for it. Domitex, 1997

Left
Lamp family of mat glass and mat-brushed steel: floor, wall and table lamps in two sizes: »Screen«, Anta, 1998

Here, the »Habitus« lounge chair, the chest of drawers »Parallele«, his »Incognito« sofa and »Presto« table of recent years formed a harmonious ensemble with »Top«, »Swing«, »Sendai« and »Graphic«. The multiple-use functionality of the new pieces is Maly's answer to changed forms of living – and therefore to changed needs: it is flexibility that counts. Thus, »Swing« is not only an easy-chair with foot stool but also comes as a dining-room chair, and the »Top« shelf can be extended almost indefinitely. It is composed of shelf boards and simple rectangular modules. Doors can be integrated on request.

Maly's entire scenario demonstrates the potential diversity of a stylistically homogeneous collection and shows that there is a constant stream of new things that satisfy demands which have not been satisfied in this way before. »A designer has to recognize this kind of potential,« says Peter Maly, »he has to sense how people secretly wish to live.«

Sometimes, Maly also fulfills his own wishes – for example, with the chimney hooks, »conmoto«. He had always looked in vain for something like this in the shops. But the normal motive for his design work is not to make something for himself. He works on interior design ideas for people he likes, for people who love all things simple and want to feel at ease and at home. People who appreciate the »Alpha« chair (Tonon), for example, because from the front it looks like a solid »A« but reveals a dynamic curve in the backrest when viewed from the side. The legs taper from seat to floor, which makes for a contrast to the front view. Maly states: »It is boring when a chair looks as angular from the side as it does from the back.« And he has a lot more to say about the sculptural quality of a chair.

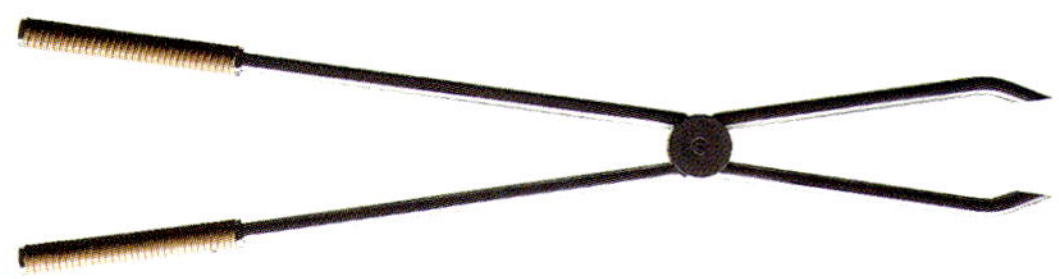

Oben
Schlichter geht's nicht: Das Kaminbesteck »conmoto« kann an der Wand hängen oder aufgestellt werden – und nirgendwo liegt rohes Holz so elegant wie auf der abnehmbaren Holzlege. Conmoto, 1997

Unten
Schreibtischsystem von großer Leichtigkeit. »Modul 5000«. Mauser Office, 1998

Above
Utmost simplicity: the chimney hooks »conmoto« can be fixed to the wall or stand on the floor. And never have rough logs found a more elegant rest than the removable rack. Conmoto, 1997

Below
Extremely light-weight desk system. »Modul 5000«. Mauser Office, 1998

Linke Seite
Bühne für die »Peter Maly Collection«: Der ligne roset-Messestand auf der Möbelmesse in Köln.

Rechte Seite
Im denkmalgeschützten Hamburger Stilwerk richtete Maly eine moderne Verkaufs- und Ausstellungsfläche für ligne roset ein.

Skizzen 1/2 oben
Der Showroom in New York orientiert sich an bestehender Architektur: Eichenparkett und Säulenreihe dominieren auch das moderne Interieur.

Skizzen 3/4 unten
In Moskau teilt eine gläserne Treppe den sechs Meter hohen und von typischen Doppelunterzügen getragenen Raum in zwei Etagen.

Left side
Setting for the »Peter Maly Collection«: the ligne roset stand at the Cologne furniture fair.

Right side
Maly created a modern sales and showroom for ligne roset in the listed Hamburg Stilwerk building.

Drawings 1/2, top
The New York showroom is modeled on the surrounding architecture. Oak parquetry floor and row of columns also dominate this modern interior.

Drawings 3/4, below
In Moscow a glass stair gives access to the second level, inserted in the six meter high hall with its typical double joists visible on the ceiling.

Um Inszenierungen weit größeren Ausmaßes geht es bei Malys vielfältigen Präsentationsflächen (siehe: S. 100 – 117). Übernahm er bereits etliche Male die Gestaltung des Kölner Messestandes von ligne roset, so konnte er das Unternehmen 1996 mit der Einrichtung eines modernen Showrooms im denkmalgeschützten Hamburger Stilwerk überzeugen: Er wurde sogleich mit der Gestaltung weiterer neuer Showrooms betraut: unter anderem in Moskau und New York, Fertigstellung: 1998.
In jeder der Metropolen läßt sich Maly auf die bestehende Architektur ein: »Man soll ja schließlich merken, ob man in New York oder in Moskau ist.« In New York heißt das: Die Fassade eines ehemaligen Galeriehauses in SoHo wird einfühlsam überarbeitet und tiefblau gestrichen – passend zum tiefen Grün vieler Häuser in dieser Gegend und passend zum Logo der Firma. Im Innern dominieren alte Eichenholzdielen und eine Mittelachse aus dicken Säulen den knapp 30 Meter tiefen Raum, der von halbdurchsichtigen Stellwänden aus Eichenlatten strukturiert wird.
Im Moskauer Showroom fallen die für stalinistische Monumentalbauten typischen Doppelunterzüge unter der Decke auf, die Maly so bestehen ließ. Den sechs Meter hohen Raum teilte er in zwei Ebenen und verband diese mit einer Treppe aus mattem Glas. Solche Details moderner Innenarchitektur zeigen, daß Maly sowohl Respekt vor der Geschichte hat als auch die Fähigkeit, eine »moderne Antwort« auf das Bestehende zu geben.

»Unlike a cupboard, a chair has to look good from all sides,« he says, and he emphasizes that it is therefore extremely difficult to design a perfect chair – quite apart from accounting for the various body shapes and daily stress such a piece of furniture has to take. Maly, however, always knows how to closely link functional and aesthetic aspects. For his »Chess« bed (Domitex) he simply fixed a textile bed head to the wall instead of to the beechwood bed frame. A solid solution. The soft fabric (designed by Maly together with the bed linen) dominates the bed as a kind of wall hanging and opens up possibilities for the creation of new »bed scenarios.«

Maly has staged numerous far larger »scenarios« with his showrooms and exhibition spaces (see pp. 100–117). At different times before 1996 he created ligne roset's Cologne Furniture Fair stand, and in that same year he convinced the firm with the design of a modern showroom in the listed Hamburg Stilwerk building and was immediately commissioned to design further ligne roset showrooms – in Moscow and New York, among others – for completion in 1998.

In each metropolis, Maly responds to the existing urban architecture because, after all, »people should recognize whether they are in New York or in Moscow.« In New York this means: the façade of a former gallery building in SoHo will be sensitively remodelled and painted a deep blue and thus fits in both with the dark green of many houses in this area and with the firm's logo.

In the Moscow showroom Maly left the highly visible double joists under the ceiling that are so typical for the city's landmark buildings from the Stalinist period. He inserted a second level inside the six meter high hall, accessible via a mat-glass stair. These details of modern architecture show that Maly respects history but is also capable of giving a »modern answer« to a historic context.

Eine seiner wichtigsten Antworten auf bestehende Wohnprobleme ist allerdings das eingangs erwähnte Sofa »Circum« (COR). Ein neuer Drehmechanismus unter den beiden Sitzpolstern macht es möglich, diese einfach nach vorn zu schwenken und als Beinablage zu nutzen. Um diese Beweglichkeit der Sitzpolster zu erreichen, entwickelte Maly eine völlig neue Konstruktion: Die Rückenlehne mußte direkt mit dem Untergestell verbunden werden, statt sie auf der Sitzfläche zu lagern. Es wird dabei deutlich, wie die Funktion die Form bestimmt. »Circum« ist ein stationäres Möbel, das seinen Platz etwa vor dem Fernseher einnimmt und für das Maly ein Kontrastmöbel gleich mitentworfen hat: den runden Drehsessel »Circo«.
Über dieses leichte Möbel freut sich Maly besonders, denn ein Sessel kann seiner Meinung nach nicht aussehen wie ein Sofa. »Der muß beweglich sein. Damit muß ich mich meinen Gesprächspartnern zuwenden können und vieles mehr.« Er wettert: »Konventionelle schwere Sitzgarnituren sind ein Greuel für mich.« Sessel »Circo« und Sofa »Circum« wurden 1998 zum riesigen Messeerfolg in Köln. Daß es ein Einfall von Malys Frau war, der den Sessel des Erfolgsduos zum Spitzenreiter machte, zeigt wie sicher oft emotionale Entscheidungen treffen: Sie schlug eine Sesselvariante mit Baumwollgurten vor – eine Idee, die anfangs keinen der Marketingexperten überzeugte. Es zeigte sich schließlich einmal mehr, daß sich Erfolgsmöbel schwer planen lassen. Da zahlt sich ein guter Designer aus, denn der kann, wie Maly sagt, »Wohnwünsche erspüren.«

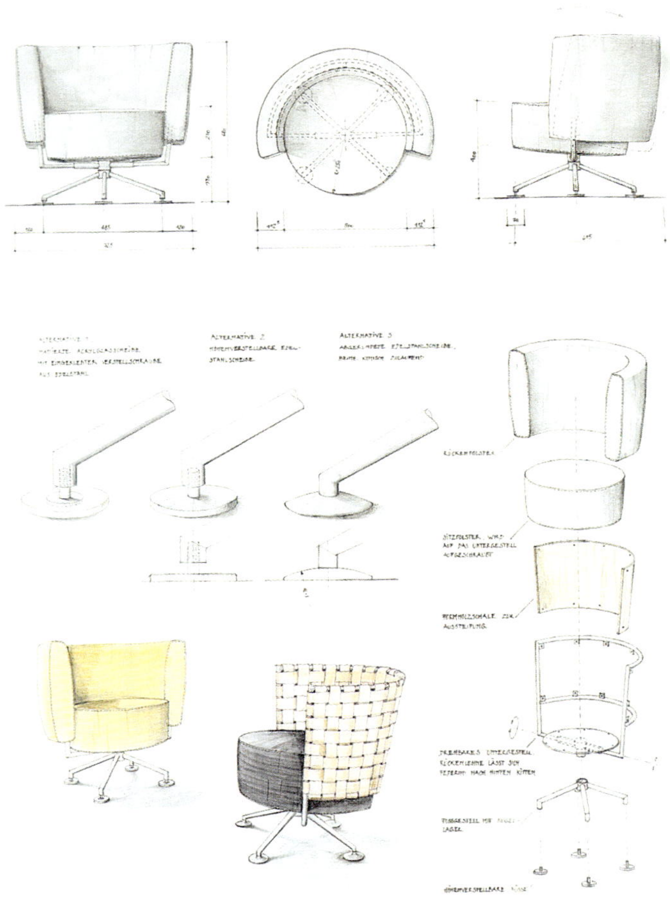

Oben und Mitte
Die zweigeteilte Sitzfläche des Sofas »Circum« läßt sich dank eines neuen Mechanismus nach vorn drehen – die elegante Art, Füße hochzulegen. COR, 1998

Unten
Der Sessel »Circo« ist das dazugehörige Kontrastmöbel. Hier die Vollpolster-Version. COR, 1998

Links
Die Zeichnung demonstriert die geometrische Präzision des Sessels »Circo«. COR, 1998

Top and middle
The two-part seat area of the »Circum« sofa can be swiveled out front thanks to a new revolving mechanism: an elegant way of putting one's feet up. COR, 1998

Below
The »Circo« armchair provides the appropriate contrast to »Circum«, the sofa. COR 1998

Left
The drawing shows the geometric precision of the covered armchair »Circo«. COR, 1998

One of his most important answers to existing problems of home décor is the already mentioned sofa »Circum« (COR). A novel revolving mechanism underneath both seat cushions makes it possible to swing them out and use them as leg rests. In order to achieve this flexibility of the seat cushions Maly developed a completely new construction: the backrest had to be directly connected to the supporting frame instead of being supported by the seats. This shows how form follows function. »Circum« is a stationary piece of furniture that takes its place in front of the television set, for example. Maly complemented it with a contrasting piece: the round revolving lounge chair »Circo«.

Maly is especially happy with this light piece of furniture because he feels an easy chair can't look like a narrow sofa. »It has to be flexible. It has to enable me to face the people I am talking to.« And he adds vehemently: »To me conventional heavy sofas and armchairs are a horror.« At the 1998 Cologne Furniture Fair, his »Circo« revolving chair and »Circum« sofa were a huge success. The fact that the success duo became a bestseller as a result of an idea of Mrs. Maly shows how sound and to the point »gut decisions« can be. She suggested a chair version with cotton webbing – an idea that initially did not convince any of the marketing experts. This goes to show once again that best-selling successes are hard to plan. So it is worth employing a good designer since he knows how to »pick up home décor desires intuitively,« says Maly.

Solange ich denken kann, besteht die Zusammenarbeit zwischen Peter Maly und COR. Als kleiner Junge war ich von seinen Möbeln und Messeständen einfach nur beeindruckt. Im Laufe der Jahre und vor allem seit ich selbst mit ihm zusammenarbeite, hat sich mir die Persönlichkeit dieses Designers nach und nach erschlossen: Neben seiner Kreativität sind die außergewöhnliche Sensibilität und der Perfektionismus, mit denen Peter Maly ans Werk geht, als Gründe für seinen Erfolg zu nennen. Hinzu kommt, daß er niemals stehengeblieben ist, sondern neue Strömungen offen aufnimmt und auf die ihm eigene Weise verarbeitet. Deshalb wird Peter Maly auch in Zukunft wichtige Beiträge für zeitgemäßes Design liefern.

Peter Maly and COR have been a team as long as I can remember. As a young boy I was just impressed by his furniture and trade fair showrooms. Over the years, and especially since I have been working with him myself, I have come to understand his designer personality more intimately. Next to his creativity it is Peter Maly's extraordinary sensitivity and perfectionism, applied to any given design task, that make for his great success. And he has never got into a fixed routine, but has always been open to new trends and ideas, which he has incorporated in his designs in his own inimitable way. This is why, in future, Peter Maly will continue to make important contributions to contemporary furniture design.

Leo Lübke

Geschäftsführer
COR Sitzmöbel,
Rheda-Wiedenbrück

Managing Director
COR Sitzmöbel,
Rheda-Wiedenbrück

Rauminszenierung im Fotostudio: Sofa »Circum« und Sessel »Circo«. COR, 1998

Space scenario in the photographer's studio: sofa »Circum« and armchair »Circo«, COR, 1998

Die gurtbespannte Version des Sessels »Circo« hatte den größten Erfolg auf der Kölner Möbelmesse. COR, 1998

The webbing version of the »Circo« armchair was highly successful at the Cologne Furniture Fair. COR, 1998

Anhang Appendix

Biographie

1936
Peter Maly wird am 19.8.1936 als ältester Sohn der Familie Maly in Trautenau (heute Trutnov/Tschechien) geboren. Dort verbringt er seine Kindheit sowie die ersten Schuljahre. Die Eltern betreiben gemeinsam ein großes Schneideratelier. Der Krieg verändert alles: 1945 fällt der Vater an der Front.

1946
1946 erfolgt die Vertreibung der Familie aus Böhmen. In der hessischen Kleinstadt Herborn besucht Peter Maly die Realschule, wo sein Zeichenlehrer das künstlerische Talent des Jungen entdeckt. Aus dieser Zeit rührt der – von der Mutter unterstützte – Entschluß, einen gestalterischen Beruf zu ergreifen.

1954
Peter Maly absolviert eine Tischlerlehre, die er 1957 beendet. Sein Gesellenstück ist eine selbstentworfene Anrichte aus Ulmenholz.

1957
Nach der Gesellenprüfung beginnt er ein Innenarchitekturstudium an der Fachhochschule Detmold, das er 1960 abschließt.

1960
Die Redaktion der Zeitschrift »Schöner Wohnen« holt den jungen Innenarchitekten nach Hamburg, wo er bis 1970 der Redaktion angehört. Hier realisiert und inszeniert er Wohnthemen und übernimmt später die Leitung des Innenarchitektenteams.

1964
1964 heiratet er Carola Ewaldsen; mit ihr hat er einen Sohn, Justus Maly. Parallel zur Redaktionsarbeit entstehen erste Designarbeiten für Tecta und COR.

1970
Peter Maly wagt den Schritt in die Selbständigkeit und eröffnet ein eigenes Studio für Design und Innenarchitektur in Hamburg. Er konzipiert weiterhin Wohnthemen für »Schöner Wohnen«, jetzt in freier Mitarbeit. Außerdem entwickelt er Fotokonzeptionen für die Möbelindustrie und erarbeitet Anzeigenkampagnen.

1971
Die langjährige Zusammenarbeit mit Interlübke nimmt mit der innenarchitektonischen Konzeption von Anzeigen und Katalogfotos ihren Anfang. Messestände, Ausstellungsräume und Möbelprogramme folgen. Peter Maly entwickelt Textilkollektionen und wird künstlerischer Berater des Unternehmens.

1974
Peter Maly heiratet nach der Scheidung seiner ersten Ehe die Stylistin Gisela Harms, mit der er bis heute auch beruflich eng zusammenarbeitet und die gemeinsame Tochter Laura hat.

1975
Für COR konzipiert er neben Möbeln auch Messestände und gestaltet firmeneigene Showrooms.

1978
Der Innenarchitekt entwickelt erste Entwürfe für Reim Interline.

1982
Mit seiner Arbeit für die französische Firma Ligne Roset beginnt der Designer auch für Unternehmen im europäischen Ausland tätig zu werden. Maly konzipiert Möbel, Messestände sowie das Konzept für eine Ligne Roset-Studiokette.

1984
Die »Schöner Wohnen«-Redaktion kürt Malys Sessel »Zyklus« (COR) zum Möbel des Jahres 1984; seine Sitzgruppe »Impuls« (W. Knoll) wird vom Designcenter Stuttgart mit der Aufnahme in die Deutsche Auswahl ausgezeichnet.

1985/1986
Ein zweites Mal prämiert das Designcenter Stuttgart einen Entwurf des Hamburger Designers: den Konsoltisch »Tabula« (Reim Interline). Mit diesem Produkt ist Maly auch einer der Preisträger des Wettbewerbs »Made in Germany«.
Das folgende Jahr bringt zwei weitere Prämierungen: Den Bundespreis »Gute Form« für seine Textilkollektion »Graphis« (Interlübke) sowie den »Du Pont International Award« für die Couch »Prao« (Ligne Roset).

1987
Von 1987 bis 1989 gestaltet Peter Maly für Mercedes-Benz mehrere Messestände in Frankfurt am Main, Paris und Genf.
Der Wahl-Hamburger wird als einer der »10 Top-Designer Deutschlands« mit dem »Gianfranco-Ferre-Preis« ausgezeichnet und sein Sessel »Zyklus« innerhalb der Ausstellung »Design Dasein« im Hamburger Museum für Kunst und Gewerbe gezeigt.

1988
Das Haus Industrieform in Essen zeichnet Peter Malys Einzelsessel »Logo« (COR) aus.

1989
Prämierungen auch im nächsten Jahr: »Hommage de l´Industrie aux Createurs« in Paris für das »Peter Maly Bett« (Ligne Roset) und der Grand Prix de la Critique, Paris, für die Recamiere »Antares«, ebenfalls für Ligne Roset. Luminaire in Miami/USA zeigt Malys Arbeiten für Interlübke im Rahmen einer Sonderausstellung über europäische Designer.

1990
Der Designer beginnt seine Zusammenarbeit mit italienischen Herstellern wie Brunati, Bros's, Consonni und Poliform sowie der Schweizer Firma Intra Design.

1991
Eine umfassende Zusammenarbeit mit der Möbelfabrik Behr beginnt.

1993
Für JAB-Anstoetz entsteht die Teppichkollektion »Peter Maly Edition«. Der Designer ist darüber hinaus für die Foto- und Katalogkonzeption verantwortlich.
Im Rahmen der Hamburger Designkreuzzüge sind diese Teppiche Gegenstand der Einzelausstellung »Flying Carpets«.
Peter Maly ist bei der »Sedia Europea Made in Italy« Teilnehmer für die Bundesrepublik Deutschland.

1994
Maly erhält die Auszeichnung »Hommage du Salon du Meuble de Paris«, und seine Objekte sind Gegenstand einer Einzelausstellung im Berliner Roland Forum. Der Designer entwirft nun auch für Thonet und den italienischen Hersteller Tonon. Anläßlich des zehnjährigen Jubiläums des »Zyklus« erscheint ein »Buch zum Möbel« sowie eine auf 500 Stück limitierte und signierte Sonderedition des Sessels.

1995
Das Design Zentrum Nordrhein Westfalen vergibt den »Roten Punkt« für hohe Designqualität an Malys Stuhl- und Tischfamilie »Thonet 737« (Thonet) sowie an sein Regalsystem »Basic« (Reim Interline).
Das Buch »Designermonographie 5 – Peter Maly« erscheint im Verlag form.
Entwurf der Sitzmöbel »Habitus«, »Incognito«, der Einzelteppiche »Les Points« und des Programmes »Parallele« für ligne roset.

1996

Peter Maly erhält die Auszeichnung für gutes Design vom Industrieforum Design, Hannover, für Stuhl und Tisch »737«/»1737« (Thonet).
Planung mehrerer ligne roset-Studios im »Stilwerk«/Hamburg, in Leipzig und Moskau.
Entwurf des Systemmöbel-Programms »Menos« für Behr.
Entwurf des Sitzmöbel-Programms »Mato« für COR.
Die Zusammenarbeit mit dem Piano-Hersteller Sauter beginnt. Entwurf der Pianos »Imago«, »Artes« und »Cura«.
Eine Einzelausstellung findet in Konstanz mit den Pianos und anderen Designarbeiten statt.
Re-Design des Designklassikers »duo« (Systemmöbel) für Interlübke.
Entwurf der Stuhlfamilie »Rota« für den italienischen Hersteller Tonon.
Einzelausstellung von Peter Malys Design-Arbeiten in Amsterdam.

1997

Die neue »Collection Peter Maly«, hergestellt von ligne roset, Frankreich, wird vorgestellt.
Re-Design des »System«-Programms – jetzt »Tecton« – für behr.
Entwurf des Flügels »Vivace« für den Pianohersteller Sauter.
Entwurf des Schlafzimmer-Programmes »Chess« für Domitex.
Einzelausstellung von Malys Design-Arbeiten in Wien.
Auszeichnungen: Design Preis Schweiz und internationaler Designpreis des Landes Baden-Württemberg für »Menos« (behr).
Entwurf von Kaminzubehör (Kaminbesteck und Holzlege) für die Firma Conmoto.
Entwurf der Stuhl- und Tischfamilie »Alpha« für Tonon.
Neue Einzelteppich-Entwürfe »Les Carrées« für ligne roset.

1997-1998

Entwurf mehrerer ligne roset-Studios in Berlin, Moskau, St. Petersburg und New York.

1998

Neue Designarbeiten für COR: Einzelsessel »Circo« und Funktionssofa »Circum«.
Bettwäsche-Kollektion für die Firma Domitex.
Entwurf eines neuen Pianos »Onda« für die Sauter Pianofortemanufaktur.
Entwurf der Leuchtenfamilie »Screen« für Anta.
Auszeichnungen: »Exhibition Design Award« (2x Silber) für die Messestände der Firma COR und ligne roset auf der IMM, Köln.
Peter Maly wird vom Industrieforum Hannover für das Kaminbesteck und die Holzlege der Firma Conmoto ausgezeichnet.
Auszeichnung vom Design Zentrum Nordrhein Westfalen: Roter Punkt für hohe Designqualität für den Flügel »Vivace« und das Funktionssofa »Circum«.
Einzelausstellung aller Pianos und des Flügels »Vivace« bei Beethoven Pianos in New York.
Zur Zeit arbeitet Peter Maly an neuen Möben für COR und Behr sowie an einem Büromöbelsystem für Mauser Office, desweiteren an neuen Ligne Roset-Studios in Hamburg und Berlin. Er konzipiert neue Messestände für die Firmen COR und Ligne Roset für die Internationale Möbelmesse in Köln.

Peter Maly lebt zusammen mit seiner Frau und seiner Tochter in Hamburg.

Biography

1936
Peter Maly was born on August 19, 1936 in Trautenau (now Trutnov in the Czech Republic), the eldest son of the Maly family. Here he spent his early childhood, as well as his first years at school. The war changed everything: his father died on the Front in 1945.

1946
The family was driven from their Bohemian homeland. Peter Maly attended secondary school in the Hessian town of Herborn, where a teacher discovered the boy's talent for drawing. Maly resolves to aspire to a creative profession, a decision which his mother supported.

1954
Peter Maly began an apprenticeship in cabinetmaking, which he completed in 1957. He designed an elmwood sideboard as his final examination project.

1957
After completing his apprenticeship, he began studying interior design at the Fachhochschule Detmold, graduating in 1960.

1960
A position on the editorial staff of »Schöner Wohnen«, a German interior decor journal, took the young interior designer to Hamburg. He worked there until 1970, realizing and presenting interior design themes. He was later put in charge of the interior design team.

1964
He married Carola Ewaldsen; they had a son, Justus Maly. Parallel to his work for »Schöner Wohnen«, Peter Maly produced his first designs for Tecta and COR at this time.

1970
Peter Maly took a first step toward independence and opened his own studio for design and interior design in Hamburg. He continued to design interiors for »Schöner Wohnen«, although now on a free-lance basis. In addition, he developed photo concepts for the furniture industry and devised ad campaigns.

1971
His collaboration of many years with the Interlübke company began with interior design concepts for ads and catalogue photos. These were followed by designs for trade fair stands, exhibition halls and furniture programs. Maly designed fabric collections and became the company's artistic consultant.

1974
Following his divorce, Peter Maly married stylist Gisela Harms; later, their daughter Laura was born. The two also enjoy a close professional relationship that has lasted up until the present.

1975
Peter Maly designed furniture, trade fair stands, and showrooms for COR.

1978
Peter Maly devised his first designs for Reim Interline.

1982
Maly branched out to companies in other European countries, starting with the French firm Ligne Roset. Peter Maly designed furniture and trade fair stands for this firm, and developed the concept for a chain of Ligne Roset studios.

1984
»Schöner Wohnen« elected Maly's »Zyklus« armchair (COR) »Furniture Item of the Year« for 1984. Stuttgart Design Center honoured his »Impuls« seating group (W. Knoll) by including it in the Final Selection of German Furniture

1985/1986
The Stuttgart Design Center again selected one of Maly's pieces for an award: the »Tabula« console table (Reim Interline). Maly also placed in the »Made in Germany« competition with this product.
Maly's work received two further awards the following year: the nationwide »Good Form« prize for his »Graphis« textile collection (Interlübke), and the »Du Pont International Award« for his »Prao« couch (Ligne Roset).

1987
Between 1987 and 1989, he designed trade fair stands for Mercedes-Benz in Frankfurt am Main, Paris and Geneva. Now by choice a resident of Hamburg, he was distinguished as one of »Germany's Top-Ten Designers« and awarded the »Gianfranco-Ferre« prize. His »Zyklus« chair was shown in the Museum for Arts and Crafts in Hamburg as part of the »Design Presence« exhibition.

1988
Peter Maly's »Logo« armchair (COR) received a prize from the Industrieform company in Essen.

1989
The following year also brought its share of prizes: »Hommage de l'Industrie aux Createurs« in Paris for the »Peter Maly Bed« (Ligne Roset) and the Grand Prix de la Critique, Paris, for the »Antares« day bed, also designed for Ligne Roset.
Maly's work for Interlübke was displayed at Luminaire in Miami, USA, as part of a special exhibition on European designers.

1990
Maly started working together with Italian manufacturers such as Brunati, Bros's, Consonni and Poliform, as well as the Swiss company Intra Design.

1991
Maly started working on a wide scale with the furniture manufacturer Behr.

1993
The rug collection dubbed »Peter Maly Edition« for JAB Anstoetz appeared. Furthermore, Maly was put in charge of the photo and catalogue concept for the company. As part of the Hamburg »Design Crusades«, these rugs were the focus of »Flying Carpets«, a solo exhibition. Peter Maly represented Germany in »Sedia Europea Made in Italy«.

1994
Maly received the »Hommage du Salon du Meuble de Paris« award, and his designs went on display in a solo exhibition in Berlin's Roland Forum. The designer put his ideas to work for Thonet and Italian producer Tonon. A book documenting »Zyklus« was published to occasion the 10th anniversary of the chair, which was produced in a special signed, limited 500-piece edition.

1995
Maly's »Thonet 737« table and chair series (Thonet), as well as the »Basic« shelf system (Reim Interline) received the »Roten Punkt«, an award for exceptional design quality, from the Design Center of North Rhine-Westphalia.
Publication of the book, »Designermonographie 5/ Peter Maly« by Verlag form. Design of the chairs »Habitus«, »Incognito«, the one-off carpets »Les Points« and the furniture line, »Parallele« for ligne roset.

1996

Awarded distinction for good design by Industrie Forum Design, Hanover, for the chair and table »737«/ »1737« (Thonet).
Designs for several ligne roset studios: in Hamburg (»Stilwerk«), Leipzig and Moscow.
Designed the systems furniture programme, »Mato«, for COR.
Peter Maly started cooperation with piano manufacturer Sauter, designed the »Imago«, »Artes« and »Cura« pianos.
Solo exhibition in Constance with pianos and other designs.
Re-design of the classic, »duo« (furniture system), for Interlübke.
Design of the chair family, »Rota«, for the Italian firm Tonon.
Solo exhibition in Amsterdam.

1997

Launching of »Collection Peter Maly«, produced by roset, France.
Re-design of the »systems« programme – now called »Tecton« – for behr.
Peter Maly designed the grand piano, »Vivace«, for Sauter, and the bedroom furniture line, »Chess«, for Domitex.
Solo exhibition in Vienna.
Maly received the Design Award Switzerland and International Design Award of Baden-Württemberg for »Menos« (Behr).
Design of the chimney hooks and log-rest for Conmoto as well as of the chairs and tables, »Alpha«, for Tonon.
New one-off carpet designs, »Les Carrées«, for ligne roset.

1997-98

Peter Maly designed and executed several ligne roset studios: in Berlin, Moscow, St. Petersburg and New York.

1998

New design commissions from COR: one-off armchair, »Circo«, and functional sofa, »Circum«.
Bed-linen collection for Domitex.
Design of the piano, »Onda«, for Sauter as well as of the lamp family, »Screen«, for Anta.
Maly received two silver »Exhibition Design Awards« for the trade fair stands of COR and ligne roset.
Distinguished by Industrie Forum Design, Hanover, for chimney hooks and log-rest (Conmoto).
Distinguished by Design Center of North Rhine-Westphalia: »Roter Punkt« for high design quality of »Vivace« grand piano and »Circum« functional sofa.
Exhibition of all Maly's pianos and his »Vivace« grand piano at Beethoven Pianos, New York.
Works on new furniture for COR and Behr; on an office furniture system for Mauser Office and new ligne roset studios in Hamburg and Berlin. Maly also designs new trade fair stands for COR and ligne roset for the Cologne International Furniture Fair.

Peter Maly lives in Hamburg with his wife and his daughter.

Bibliographie Bibliography

Unsere Auswahlbibliographie gibt einen Überblick über die wichtigsten Zeitschriften- und Zeitungsartikel der letzten Jahre.

Our select bibliography provides an overview of the most important newspaper and magazine articles of recent years.

Interviews mit Peter Maly

Interviews with Peter Maly

Katrin Ambühl: *Mobilität sollte nicht zur modischen Attitude verkommen.* In: Wohnrevue 1/1998, S./p. 42
Carsten Dürer: *Vom Sessel zum Flügel.* In: Piano News 1/1997, S./p. 24-27
Robert Haidinger: *Der lobenswert unbunte Hund.* In: Wohnen 6/94, S./p. 60-64
Rita Henss: *Fragebogen.* In: F.A.Z.-Magazin vom 28.10.1994, S./p. 42
Interview. In: ARD – Hier und Heute. Sendung anläßlich der Möbelmesse Köln 1995 auf dem COR-Messestand am 20. Januar 1995
Interview und Filmbeitrag. In: HH1 – Zu Gast bei ..., 7.1.1996
Interview und Filmbeitrag. In: ZDF-Mittagsmagazin am 17.1.1995
Möbeldesign. In: form. Zeitschrift für Gestaltung 1/1992, S./p. 55
Peter Maly – das DDS-Messegespräch in Köln. In: DDS/Der deutsche Schreiner + Tischler 4/1993, S./p. 26f
Evelyn Preuss: *Ich will keine Wegwerfmöbel.* In: Hamburger Abendblatt, Extra Journal vom 24.4.1992, S./p. 2
Birgit Röthel: *Gutes Design markiert Lebensspuren.* In: Polster-Fashion 2/1991, S./p. 34-38

Porträts über Peter Maly

Portraits of Peter Maly

Rolf Brockschmidt: *Das Sofa zum großen Gatsby.* In: Der Tagesspiegel vom 20.4.1994, S./p. 16
Dieter Bross: *Peter Maly – von zeitloser Eleganz.* In: Menstyle Frühjahr/Sommer 1992, S./p. 92-96
Christina Buse: *Das Geheimnis seines Erfolges.* In: Schöner Wohnen 2/1993, S./p. 98-104
Josette Cagli: *Werkstattgespräch: Design für den Alltag.* In: Wohn-Idee 5/1988, S./p. 105-108
Design-Bilder. In: md 7/1992, S./p. 58-60
Pierre Darge: *Zoeken naar vanzelfssprekenheid.* In: Weekend-Knack 36/1995, S./p. 262-266
Robert Doerk: *Die Deutschen kommen.* In: Ambiente 11/1989, S./p. 168
Ein opulenter Minimalist. In: Domizil 4/1996, S./p. 9-15
Jan Esche: *Peter Maly oder die Liebe zur Geometrie.* In: Design Report 1/1994, S./p. 66-69
Anna Fadeeba: *Peter Maly – Tischler, Journalist, Designer.* In: Salon Interieur 6/96 (Moskau), S./p. 84-87
Lambert Fidel: *Stühle, auf denen man sitzen will.* In: Möbel Raum Design 10/11/1996, S./p. 46-52
Kurt Gustmann: *Peter Maly oder die Liebe zur klaren Form.* In: Schöner Wohnen 11/1985, S./p. 253-261
Wolfgang Hirn: *Ein Grüner in Schwarz.* In: Manager Magazin 11/1987, S./p. 370-375
M. Huthmacher: *Plädoyer für Disziplin.* In: Arcade 1/1991, S./p. 12
Corine Koole: *Het Huis van Peter Maly als decor voor zijn tapijten.* In: Eigenhuis & Interieur 10/1994, S./p. 113ff
Lore Kelly: *Aus Liebe zur Perfektion.* In: Raum und Wohnen 4/1991, S./p. 90-96
Lore Kelly: *Peter Maly – geometrischer Konstruktivist.* In: Wohnrevue 5/1990, S./p. 107-117
Martin Krautter: *Designerportrait über Peter Maly.* In: F.A.Z.-Beilage »Design« vom 29.11.1994, S./p. B 12
K. Leydecker: *Designer-Portrait Peter Maly.* In: Möbel Raum Design 2/1994, Titelfoto u. S./p. 86-101
Birgit Nordmann: *Design-Superstar Peter Maly.* In: Atrium 2/1996, S./p. 100-107
Peter Maly – Designerportrait. In: Inspiration, International Design Manual 1992, S./p. 2-5
Peter Maly – le design par excellence. In: Maison Montreal 1/1989, S./p. 78f
Jens Priewe: *Der Mann, der mit gutem Geschmack die Welt verändern will.* In: Ambiente 3/1981, S./p. 122-135
D. Reinbender: *Peter Maly – Designerportrait.* In: Möbelkultur (Messeausgabe) 1/1977, S./p. 166-171
Gisela Reiners: *Hat Erfolg in Italien mit deutschen Ideen.* In: Die Welt, Hamburg-Lifestyle, 8.8.98, S./p. 25
Regine Scourtelis: *Ein Realist mit Eigensinn.* In: Elle Decoration 22/1995, S./p. 39-43
Susanne Schwarz: *Design-Geschichte in Jahrzehnten – Peter Maly.* In: Privé 10/1995, S./p. 102-111
Judka Strittmatter: *Möbel sind keine modischen Fummel.* In: Welt am Sonntag Nr. 36, S./p. 29
Desirée Stürgkh: *Mann mit FormArt.* In: H.O.M.E. 3/1997, S./p. 30-36
Cristina Thies: *Harmonie der Gegensätze.* In: Schöner Wohnen 11/1995, S./p. 43-50ll
Hans Joachim Weimann: *Möbel von Vererbungsqualität.* In: Hamburger Abendblatt vom Juni 1983
Dr. Peter Zec: *Portrait Peter Maly.* In: German Design Standards 1998, S./p. 114-117

Veröffentlichungen zu Produkten

Publications on Maly's product designs

Agora – *65 oder 130 Zentimeter.* In: md 5/1990, S./p. 72ff
Alpha, Beta, Gamma – md 1/1981, S./p. 30-32
Art Deco – *Art Deco für die achtziger Jahre.* In: Architektur & Wohnen 3/1980, S./p. 158-162 – md 10/1980, S./p. 76-79
Artes, Cura, Imago, Vivace. In: Monthly Design, Seoul/Korea 22/1997
Axis – *Trennt und verbindet – das Paneel.* In: md 1/1990, S./p. 62f
Cena, Mato, Rota, Sera, Thonet. In: Ottogano 118, S./p. 130
Cena – *Mix'n'Match mit starkem Rücken.* In: md 11/1996, S./p. 64/65
Circum, Circo, Ganymed, Menos. In: Zuhause Wohnen 2/1998, S./p. 24/25
Cirrus – *Ein Möbel für 24 Stunden.* In: md 3/1992, S./p. 54f
Cycla – *Senzaparole.* In: DDN/Design Defusion News 10/1994, S./p. 58f
Duo-Bar – *Ein Richtfest für eine Bar.* In: Lui 2/1988, S./p. 54
Duo-Programm – *Duo entwickelt sich.* In: md 5/1988, S./p. 44-46 – Interni 8/1989, S./p. 72 – Raum und Wohnen 2/1990, Titel u. S./p. 105
Elliptic – *Hausschränke Schrankhäuser. Architekturmöbel aus zwei Jahrhunderten 1790-1990.* Orangerie Draenert. Immenstaad/Bodensee 1991, S./p. 28f
Edition Peter Maly – *Maly-Schränke.* In: md 1/1993, S./p. 66-69
Facades – Dirk Meyhöfer: *Angeregt von Altstadtgiebeln.* In: Architektur & Wohnen 5/1992, S./p. 120-122 – *Schränke hinter Fassaden.* In: md 7/1987, S./p. 44f – *Maly-Schränke.* In: md 1/1993, S./p. 66-69
Menos – *Im Wesentlichen einfach.* In: md 9/1996, S./p. 53-55
Menos + Tecton. In: Möbelkultur 1/1998, Titel, S./p. 78/79
Metrix – *Möbel-Architektur.* In: md 11/1992, S./p. 56-59
Metrix-Wohnen – *Die Wand dazwischen.* In: Ideales Heim 4/1992, S./p. 60f
Metrix-Schlafen – *Vielfalt als Prinzip.* In: md 9/1994, S./p. 49-51
Minos – *Hausschränke Schrankhäuser. Architekturmöbel aus zwei Jahrhunderten 1790-1990.* Orangerie Draenert. Immenstaad/Bodensee 1991, S./p. 28f
Onde – *Das Prinzip: Ein Magazin.* In: md 1/1992, S./p. 60f
Peter Maly Bett – *Schöner kann Schlafen nicht werden.* In: md 11/1990, S./p. 95
Peter Maly Teppichedition – *Materialcollagen.* In: md 12/1994, S./p. 66f
Prao – *Confidents pour confidences.* In: Maison Jardin 10/1987, S./p. 148
Pub – *65 oder 130 Zentimeter.* In: md 5/1990, S./p. 72-75
Quadrat – *Möbel, die Geschichte machen/Moderne Klassiker.* Gruner + Jahr, Hamburg 1993, S./p. 99
Quartz – *Ein System mit klaren Formen.* In: Schöner Wohnen 1/1989, S./p. 28-30

Werkverzeichnis

Serie 1 und 2 – *And this you can do anywhere.* In: Abitare 11/1972, S./p. 152f
Tarbo – *Radien prägen die Form.* In: md 9/1990, S./p. 114-116
Tecton – *Freiheit im System.* In: md 11/1998, S./p. 57-59
Thonet 737 – *Stuhl der Woche.* In: Die Woche vom 9.12.1994, S./p. 62
Ticino – *Variantenreich … Ticino.* In: form. Zeitschrift für Gestaltung 2/1995, S./p. 66
Veritas – *Architektonische Reminiszenzen.* In: md 8/1988, S./p. 35-37
40 SN – *Möbel, die Geschichte machen/Moderne Klassiker.* Gruner + Jahr, Hamburg 1993, S./p. 189
Zyklus – Ausstellungskatalog *Design Dasein*, Museum für Kunst und Gewerbe Hamburg, Hamburg 1987, S./p. 55
International Design Yearbook 1986/87, John Calmann and King Ltd., London 1986, Titel und S./p. 48
10 Jahre Zyklus, Hrsg. COR, Rheda-Wiedenbrück 1994
10 Jahre Zyklus. In: md 7/1994, S./p. 27
Möbel, die Geschichte machen /-Moderne Klassiker. Gruner + Jahr, Hamburg 1993, S./p. 156

Veröffentlichungen zu Messeständen und Ausstellungen

Publications on trade fair stands and exhibitions

COR Messestand – Ingrid Wenz-Gahler: *Messestand-Design*, Verlagsanstalt Alexander Koch, Leinfelden-Echterdingen 1995, S./p. 110f
Zur Schau gestellt. In: md 6/1988, S./p. 78f
Egon Schirmbeck/Eduard Schmutz: *Messestände.* Verlagsanstalt Alexander Koch, Leinfelden-Echterdingen 1987, S./p. 28-33
COR Showroom – *Schnell am Ziel.* In: Design Report 3/1995, S./p. 96f
Interlübke Messestand – *Stand-Show für die Möbel-Show.* In: md 3/1986, S./p. 42-45
Interlübke Showroom – *Schau im Showroom.* In: md 10/1985, S./p. 70f
Salon ligne roset. In: Archidom (Moskau), S./p. 68-75

Beim nachfolgenden Werkverzeichnis haben wir uns bemüht, eine möglichst vollständige Liste aller von Peter Maly gestalteten Produkte zu erstellen. Sollten dennoch Objekte fehlen, bitten wir den Leser um Verständnis.
Die mit Stern gekennzeichneten Produkte werden nicht mehr hergestellt und sind daher im Handel nicht erhältlich.

1967 – Serie 1*
Einzelmöbelgruppe. Farbig lackiert und mit Dekor-Siebdrucken (Tecta)

1969 – Trinom*
Sitzmöbel-Elementgruppe mit Formholz-Rückenschalen (COR)

1973 – Serie 2*
Einzelmöbel-Kollektion. Lackiert (Tecta)

1975 – Serie 5*
Einzelmöbelserie. Kork und Eschenholz (Tecta)

1975 – Cento*
Sitzelemente zum Kombinieren (COR)

1976 – Delta*
Sessel und Sofa mit Aluminium-Rohrgestell (COR)

1976 – Serie 6*
Kommoden und Tische mit schachbrettartiger Furnierung, Eschenholz und Porzellanknöpfen (Tecta)

1977 – M 70
Tisch mit Aluminium-Gußgestell im Quadratraster, Glasplatte (Tecta)

1977 – Serie 7*
Kommodenserie zum Hängen und Stapeln. Geflechtfüllungen in Massivholzrahmen (Tecta)

1978 – Serie 1*
Re-Design der Einzelmöbel-Gruppe von 1967 (Tecta)

1978 – Citta*
Couch- und Sesselprogramm in prägnanter Linienführung mit zwei Rückenhöhen (COR)

1978 – Combra
Stuhl- und Tischfamilie. Eschenholz und Leder (COR)

1978 – Taxus*
Solitärmöbelkollektion. Eibenholz und Travertin (Reim Interline)

1978 – Basis*
Innenausbauprogramm mit Regalen, Paneelen und Schränken. Eibenholz und Lack (Reim Interline)

1978 – Beverly*
Couchtische in mehreren Formaten. Holz, lackiert (COR)

1978 – Lampenkollektion I*
»Alpha«, »Beta«, »Gamma«, »Delta«, »Epsilon«, »Zeta«, »Kappa«, »Eta«. Nickel, Messing, Travertin und Marmor (Reim Interline)

1979 – Mutaro*
Schlafzimmer-Ensemble: Betten, Kommoden, Beistelltische und Wandpaneele. Lack und Kirschbaum (Interlübke)

1980 – Octa* (ab 1982 »Octanova«)
Leichte Vollpolstercouch und Sessel (COR)

1980 – Quadrat
Aus dem 1978 entworfenen Programm »Taxus« hervorgegangen (überarbeitet 1993 »Quadrat Nova«). Vitrinenschränke mit Massivholzrahmen im Quadratraster (Reim Interline)

1980 – MY*
Leseleuchte. Messing vernickelt (Hersteller Anta, Vertrieb Reim Interline)

1980 – Art Deco (ab 1987 »Art Deco Nova«)*
Solitärmöbel-Programm. Vogelaugenahorn, gebeizt und Hochglanzlack, seit 1992 Kirschbaum/Mattlack-Kombination (Reim Interline)

1981 – Modus*
Kastenmöbel-Programm: Einzelschränke, Kommoden, Eßtisch. Eschenholz, offenporig lackiert (Renz)

1982 – Mistral*
Großvolumige Couch mit Eschenholz-Rahmen (Kill)

1983 – Peter Maly-Bett
Bett mit breitem gepolstertem Rahmen (seit 1994 auch mit Buchenholz-Rahmen) mit schwenkbaren Tabletts und einsteckbaren Rückenkissen (Ligne Roset, Frankreich)

1984 – Duo
Basisprogramm (gemeinsam mit Hans Hopfer). Systemmöbel für Wohn- und Schlafbereich, bestehend aus »Containern« und dazwischen gespannten Böden; Betten und Beimöbel. Lack und Buchenholz (Interlübke)

1984 – Tabula*
Konsoltische, vergrößerbar zu Eßtischen (Reim Interline)

1984 – Horizon*
Regalsystem mit waagerechter Linienführung. Hochglanzlack, schwarz und weiß (Reim Interline)

1984 – Konos*
Tischleuchte. Kegelfuß Marmor weiß oder schwarz (Reim Interline)

1984 – Zyklus
Sessel mit Stahlrohr-Armlehnen, rollbar, mit verstellbarer Nackenrolle (COR)

1984 – Variation
Regal-Anbausystem. Holz/Lack-Kombinationen (Ligne Roset, Frankreich)

1985 – Prao
Couch mit einsteckbaren Rücken und Armlehnen (Ligne Roset, Frankreich)

1985 – Graphis*
Textil-Kollektion. Baumwolle bedruckt (Interlübke)

1985 – 40 SN*
Schrankwand. Re-Design des Klassikers der Team Form von 1963 (Interlübke)

1986 – Duo
Erweiterung: Bar und Cockpit-Schreibtisch (Interlübke)

1986 – Quartz
Regal-Anbausystem und Eßtisch. Lack, Eschenholz und Metall (Ligne Roset, Frankreich)

1987 – Muro*
Paneel- und Schrank/Anbausystem (WK Wohnen)

1987 – Facades
Serie von Singulär-Schränken mit unterschiedlichen Fassaden und Schiebetüren, die sich frei in den Raum bewegen lassen: »Pallas«, »Minos«, »Luna«, »Nike«. Hochglanzlack, Vogelaugenahorn (Reim Interline)

1988 – Veritas
Einzelmöbel-Programm und Eßtische. Ahorn- und Birnbaumholz (Reim Interline)

1988 – Cubic*, Luna II*
Ergänzung zur »Facades«-Serie (Reim Interline)

1988 – Zelda
Klassische Clubsessel und Sofas (COR)

1988 – Logo
Kleiner Einzelsessel (COR)

1988 – Antares*
Couch und Recamiere. Polsterteile auf Aluminium-Gestellen (Ligne Roset, Frankreich)

1988 – Lumen*
Couchtisch mit indirekter Beleuchtung. Ahornholz, Mattglas, schwarzes Metallgestell (Ligne Roset, Frankreich)

1989 – Cristalli*
Wandfliesen-Serie für Bäder (Villeroy & Boch)

1990 – Taurus
Stuhl- und Tischfamilie. Buchenholz und Lack (COR)

1990 – Tarbo
Sitzmöbel-Elemente, deren Formsprache auf großen Radien basiert (COR)

1990 – Edition Peter Maly
Objektmöbel in limitierter Auflage. Birnbaumholz, Hochglanzlack, Ahorn, Edelstahl. Schränke »La Bandiera« und »Elliptic«; Tisch »Thor«; Sekretär »Ivory« (Intra Design, Schweiz)

1990 – Arcus
Schlafraum-Programm und Bettwäsche-Kollektion. Satin, bedruckt (Domitex)

1990 – Rondo*
Einzelschrank mit gewölbter Front. Birnbaum und Lack (Reim Interline)

1990 – Hera
Stuhl aus geformten Holzschalen auf einem Metallrohrgestell (Ligne Roset, Frankreich)

1990 – Agora*
Runder Eßtisch mit blau getönter Glasplatte, auf gewölbter Schale ruhend. Fußgestell Holz und Rundrohr Stahl, schwarz (Ligne Roset, Frankreich)

1990 – Duo*
Funktionscontainer: Kofferartig aufklappbarer Container mit Bar, Hifi/TV-Funktion. Lack und Holz-Kombination (Interlübke)

1990 – Zelda
Runde Tische mit gedrechseltem Mittelfuß aus Kirsch- und Ahornholz verleimt (COR)

1990 – Axis*
Schrank- und Paneelwandsystem (Poliform, Italien)

1990 – Contur*
Couch mit drei verschiedenen Einzelsesseln (Wittmann)

1991 – Edition Peter Maly*
»Fluctus« – Schrank mit auf der Mittelachse montierten Flügeltüren. Grau gebeizter Ahorn und Birnbaum in Streifenfurnierung. »Torii« – Lackregal mit umfassender Birnbaum-Verblendung (Intra Design, Schweiz)

1991 – Tarbo
Polsterbett mit nach unten abgeschrägtem Polsterrahmen (COR)

1991 – Onde
Schranksystem für den Schlafraum. Fronten aus verformter MDF-Platte, buchenfurniert oder lackiert. Inneneinteilung Buche (Ligne Roset, Frankreich)

1991 – Boa
Sitzmöbel-Anbausystem mit schlangenförmigen Rückenlehnen (Gelderland, Niederlande)

1991 – Koga*
Sitzmöbelelemente mit hohen Rücken- und Armlehnen (Ligne Roset, Frankreich)

1991 – Tectus
Eßzimmer-Stuhl aus massivem Kirschbaum (Bros´s, Italien)

1991 – Schach
Eßtisch mit schachbrettartiger Furnierung und Intarsien (Bros´s, Italien)

1991 – Bibo
Leseleuchte, dreh- und kippbar. Nickel, Glas (Anta)

1992 – The New Biedermeier*
Solitär-Sofa (Gelderland, Niederlande)

1992 – Felix
Sitzgruppe (Brunati, Italien)

1992 – Sharky
Einzelcouch mit zahlreichen Verstellfunktionen (Brunati, Italien)

1992 – Vega (ab 1994 »Veganova«)
Sessel und Sofa mit schwungvoller Linienführung (COR)

1992 – Metrix-Wohnen
Schrankwand-System mit vielen Kombinationsmöglichkeiten und Materialien (behr)

1992 – Arc
Leichter, komfortabler Stuhlsessel. Holzschale, Metallfußgestell, Polsterung Leder oder Alcantara (Ligne Roset, Frankreich)

1992 – Quadro, Undine, Intarsia
Ergänzung der »Facades«-Solitärschränke in Birnbaum mit verschiedenen Intarsien (Reim Interline)

1992 – Ventrio*
Sideboard in Form eines Kreissegments. Kirschbaum- und Lack-Kombination (Reim Interline)

1992 – Cirrus
Sitzliege mit schwenkbaren Lehnen. In der Sitztiefe durch Verschieben veränderbar, dann als Schlafsofa nutzbar. Metallstruktur schwarz oder alufarbig; Leder- oder Stoffpolsterung (COR)

1992 – Pando
Leichter Einzelsessel mit Metallstruktur. Alufarbig, Polsterbezüge geschliffenes Leder (COR)

1993 – Cycla
Kollektion verschiedener Tisch-Accessoires: Vasen, Schalen, Leuchter, Kaffee- und Teeservice, Tabletts, Windlicht, Menage. Silber mit Birnbaum oder Edelstahl mit Buche (Wilkens Bremer Silberwaren)

1993 – Cuneo (gemeinsam mit Christian Werner)
Einzelmöbelserie bestehend aus verschiedenen Schrankmöbeln, Regalen, Eßtisch und Stuhl. Kirschholz, teilweise massiv verarbeitet (Consonni, Italien)

1993 – Biblia
System von Bücherschränken, Tischen und Beimöbeln (Ligne Roset, Frankreich)

1993 – Papilio*
Einzelschrank. Ergänzung der »Peter Maly Edition«. Birnbaum; innen: blauer Spiegel (Intra Design, Schweiz)

1993 – Peter Maly-Teppichedition
Kollektion bestehend aus: »Ganymed«, »Gondwana«, »Galaxi«, »Crossing«, »Cubes«, »Centercourt«, »Genesis«, »Match«, »Monega«, »Medio I + II«, »Punto«, »Minipoint«, »Nirwana«, »Najade«, »Nomos«. Reine Schurwolle, handgefertigt, mit eingelassenen Intarsien in Kupfer, Edelstahl, Messing, Leder und Seide (JAB Anstoetz-Teppiche)

1993 – Quadrat Nova
Ergänzung zur »Quadrat«-Serie, Schrank- und Sideboardkombinationen. Kirschbaum-Rahmen, Korpus Lack schwarz (Reim Interline)

1994 – Mosaik
Ergänzung der »Facades«-Serie. Schrank in schachbrettartiger Furnierung und mit schwarzen und weißen Holzintarsien; Kirschbaum und Lack (Reim Interline)

1994 – Thetis
Ergänzung der »Facades«-Serie. Kirschbaum mit Metallintarsien (Reim Interline)

1994 – Ticino
Eßzimmer-Kollektion: Anrichten, Vitrinen, Eßtische, Polsterstühle. Buche, Linoleum, Textil, Mattglas (WK Wohnen)

1994 – Metrix-Schlafen
Möbelprogramm für den Schlafraum: Schranksystem mit edler Innenausstattung, mit Polsterbett, Sprossenbett, Paneelbett, Beimöbeln und textiler Ausstattung (behr)

1994 – Basic
Regalsystem mit Schiebetüren und Paneelen. Kirschbaum und Aluminium (Reim Interline)

1994 – Sera
Relax-Sessel und Couch mit veränderbaren Armlehnen. Vernickelte Stahlgestelle, Leder- und Stoff-Polsterung (COR)

1994 – Cocon*
Recamiere in ovaler Form und ovale Kirschholz-Tische (COR)

1994 – Ludwigsburg
Wandleuchte für indirektes Licht. Metall, alufarbiger Strukturlack (Anta)

1994 – Thonet 737 + 1737
Stuhl- und Tischfamilie. Buchenholz-Gestell mit Baumwollgurt-Bespannung oder Polsterung. Tischplatten Buche und Linoleum (Thonet)

1995 – Parallele
Mobile Kommoden und Schränke, Betten. Buchenholz, Lack, geriffeltes Aluminium (Ligne Roset, Frankreich)

1995 – Habitus
Einzelsessel mit gebogenen Schichtholz-Armlehnen. Buchenholz, Leder, Textil (Ligne Roset, Frankreich)

1995 – Incognito
Sofa mit Hussenbezug und sichtbaren Reißverschlüssen, auf Rollen oder Füßen (Ligne Roset, Frankreich)

1995 – Presto
Couch- und Tablettisch auf Rollen (Ligne Roset, Frankreich)

1995 – Les Points
Einzelteppiche aus Sisal, mit Edelstahlösen und Wollbüscheln (Ligne Roset, Frankreich)

1995 – Margo
Polsterbett mit ovalen Schwenktabletts aus Ahornholz und exklusiver Bettwäsche aus bedrucktem Satin (Domitex)

1995 – Cena
Stuhl- und Tischfamilie. Stühle mit verschiedenen Formen der Rückenlehne (gepolstert oder Aluminium). Untergestell Buche. Tische mit Buchen- oder Metallgestell; Platten Buche, Linoleum, Glas (Tonon, Italien)

1996 – Menos
Möbelprogramm für Wohn- und Eßbereich. Kommoden in Samtlack in verschiedenen Farben. Regale, Paneele und Tische in Ahorn oder Buche. (behr)

1996 – Mato
Modulare Sitzelemente in Form von Bänken aus Massivholz mit Gurtbespannung. Lose Sitzkissen, Armlehnen und Rückenlehnen in verschiedenen Positionen einsteckbar. Dazu Couchtische mit Steinplatten sowie Einzelsessel mit Gurtbespannung. (COR)

1996 – Tameru, Tameras, Someru
Erweiterung der P.M.-Teppich-Edition. Reine Schurwolle, handgefertigt, Edelstahl- und Kupferintarsien. (JAB Anstoetz)

1996 – Imago, Artes, Cura. Drei Pianos in zeitgemäßer Formgebung. Imago: Rahmen Birnbaum, Korpus Mattlack. Artes: Hochglanzlack schwarz mit Edelstahl-Details. Cura: Kirschbaumholz mit Edelstahl-Details. (Sauter)

1996 – Duo (Re-Design)
Überarbeitung und Ergänzung des puristischen Design-Klassikers von 1984. Neue Metallgriffe, Alu/Glasrahmentüren, Regalcontainer, Eßtisch. (Interlübke)

1996 – Rota
Stuhlfamilie, wahlweise mit Rücken in Massivholz, Formholz oder gepolsterter Ausführung. Runder Eßtisch, ausziehbar, Kirschbaum oder Buche. (Tonon, Italien)

1997 – Chess
Bett mit Ahornpfosten und textilem Wandbehang, Beimöbel Ahorn mit Bettwäsche-Collection. (Domitex)

1997 – Top
Regalsystem mit einfachster Steck-Montage. Waagerechte Linienführung, Metallteile alufarbig, Böden Buche. (ligne roset, Frankreich)

1997 – Swing
Sessel und Sofa in skulpturaler Formgebung. Metallgestelle alufarbig. (ligne roset, Frankreich)

1997 – Ellipse
Drei Leuchten mit aufknöpfbaren Polycarbonat-Schirmen. (ligne roset, Frankreich)

1997 – Tecton
Systemmöbel als Re-Design des Programmes »System«. Quadratraster-Struktur, in die verschiedene Frontelemente eingeschoben werden können. Großflächige Schiebetüren schließen Funktionsbereiche. (behr)

1997 – Vivace
Flügel in zeitgemäßer Formgebung. Hochglanzlack schwarz oder Ahornholz, mit Edelstahl-Fußgestell und -Intarsien. (Sauter)

1997 – Kaminbesteck
Verschiedene Geräte in Edelstahl mit Naturleder-Umwicklung als Griff, mit Wandhalterung oder Standhalterung. (Conmoto)

1997 – Alpha
Eßzimmer-Stühle und Tische, Massivholz Buche, Ahorn oder Kirschbaum. Tischzarge Massivholz oder Edelstahl. (Tonon, Italien)

1997/98 – Les Carrées
»Quadri«, »Bayadere«, »Sendai«, Einzelteppiche aus Schurwolle. (ligne roset, Frankreich)

1998 – Carrée
Bettwäsche-Collection, Baumwolle, verschiedene Farben, mit umlaufendem Carrée-Fries. (Domitex)

1998 – Circum
Couch mit schwenkbaren Sitzkissen, durch die Vergrößerung der Sitztiefe dann als Relax-Couch nutzbar. Metallstruktur nickelfarbig lackiert, Leder- oder Stoffbezüge. (COR)

1998 – Circo
Drehbarer, leichter runder Sessel. Sitz gepolstert, Rücken wahlweise gepolstert oder mit Gurtgeflecht. Metallstruktur nickelfarbig lackiert, Leder- oder Stoffbezüge. (COR)

1998 – Onda
Piano in minimalistischem Design. Front Wellenstruktur, Alu-Lack und Ahorn. (Sauter)

1998 – Parallele-Paneelsystem
Niedrige Holzpaneele mit waagerechten Schlitzen, in die Metallböden variabel eingesetzt werden können. Böden Alu-Lack, Paneel Buche natur. (ligne roset, Frankreich)

1998 – Screen
Leuchtenfamilie: Große und kleine Tischleuchte, Stehleuchte, Wandleuchte. Mattglas-Scheiben und Metallstruktur, Metall matt verchromt. (Anta)

1998 – Modul 5000
Schreibtischprogramm mit großer Variationsbreite. Aluminiumgußgestelle mit verschiedenen Plattformen und Materialien. (Mauser Office)

List of Works

The publishers of this book have done their utmost to provide a complete list of products designed by Peter Maly. We ask the reader to accept our apologies should any objects have been left out.
The items indicated with an asterisk are no longer manufactured, and thus are not available on the market.

1967 – Serie 1*
Series of individual serially produced furniture pieces, lacquered in various colours or with silk-screen decorations (Tecta, Germany)

1969 – Trinom*
Seating element group with backrests made of formed plywood shells (COR, Germany)

1973 – Serie 2*
Individual furniture pieces, lacquered (Tecta, Germany)

1975 – Serie 5*
Individual furniture pieces. Cork and ash (Tecta, Germany)

1975 – Cento*
Seating elements which can be combined in various ways (COR, Germany)

1976 – Delta*
Easy-chair and sofa with aluminium tube frame (COR, Germany)

1976 – Serie 6*
Chests of drawers and tables with a chequered veneer. Ash with porcelain knobs (Tecta, Germany)

1977 – M 70
Table with a cast aluminium frame in a grid pattern, glass table top (Tecta, Germany)

1977 – Serie 7*
Chests of drawers for hanging or stacking. Wicker work inlays in a solid wood frame (Tecta, Germany)

1978 – Serie 1*
Redesigned version of the 1967 series (Tecta, Germany)

1978 – Citta*
Couch and armchair program with a striking design and two different backrest heights (COR, Germany)

1978 – Combra
Table and chair group. Ash and leather (COR, Germany)

1978 – Taxus*
Unique furniture line. Yew and travertine (Reim Interline, Germany)

1978 – Basis*
Interior furniture program with shelves, panelling and cupboards. Yew and lacquer (Reim Interline, Germany)

1978 – Beverly
Coffee table in various sizes. Lacquered wood (COR, Germany)

1978 – Lamp collection I*
»Alpha«, »Beta«, »Gamma«, »Delta«, »Epsilon«, »Zeta«, »Kappa«, »Eta«. Nickel, brass, travertine and marble (Reim Interline, Germany)

1979 – Mutaro*
Bedroom ensemble: beds, chests of drawers, bedside tables and wall panelling. Cherry and lacquer (Interlübke, Germany)

1980 – Octa* (»Octanova« as of 1982)
Light fully-upholstered couch and easy-chair (COR, Germany)

1980 – Quadrat
Based on the »Taxus« line from 1978 (revised in 1993). Glass cabinets with solid wood frames in a grid pattern (Reim Interline, Germany)

1980 – MY*
Reading lamp. Nickel-plated brass (Producer Anta, distributer Reim Interline, Germany)

1980 – Art Deco (later »Art Deco Nova«)
Unique furniture program. Curled maple, stained and high gloss laquer (Reim Interline, Germany)

1981 – Modus*
Cupboard furniture system. Cupboards, chests of drawers, dining table, lacquered ash (Renz, Germany)

1982 – Mistral*
King-sized couch with an ash frame (Kill, Germany)

1983 – Peter Maly Bed
Bed with a wide upholstered frame (as of 1994, also available with a beech frame). With swivelling bedside trays and slot-in backrest cushions (Ligne Roset, France)

1984 – Duo
Basic program (designed in collaboration with Hans Hopfer). Container system for living room and bedroom. »Containers« with shelves mounted between them; beds and bedside furniture. Beech and lacquer (Interlübke, Germany)

1984 – Tabula*
Console table which can be extended to serve as a dining table (Reim Interline, Germany)

1984 – Horizon*
Shelf system with a horizontal design. High-gloss lacquer (Reim Interline, Germany)

1984 – Konos*
Table lamp. Cone-shaped base in white or black marble (Reim Interline, Germany)

1984 – Zyklus
Armchair with steel tube armrests and adjustable neckroll. On casters (COR, Germany)

1984 – Variation
Shelf unit system. Wood and lacquer combination (Ligne Roset, France)

1985 – Prao
Couch with slot-in back and armrests (Ligne Roset, France)

1985 – Graphis*
Fabric collection. Printed cotton (Interlübke, Germany)

1985 – 40 SN
Wall-to-wall cupboard unit. Redesign of the Team Form classic from 1963 (Interlübke, Germany)

1986 – Duo
Extension to the basic program: bar and cockpit desk (Interlübke, Germany)

1986 – Quartz
Shelf unit system. Ash, lacquer and metal (Ligne Roset, France)

1987 – Muro*
Panelling and shelf unit system (WK Wohnen, Germany)

1987 – Facades
Series of unique cupboards with sliding doors that slide out into the room, with various facades: »Pallas«, »Minos«, »Luna«, »Nike«. Curled maple and high gloss laquer (Reim Interline, Germany)

1988 – Veritas
Program of individual furniture pieces and dining tables. Maple and pear (Reim Interline, Germany)

1988 – Cubic*, Luna II*
Extension to the »Facades« series (Reim Interline, Germany)

1988 – Zelda
Classic club chairs and sofas (COR, Germany)

1988 – Logo
Small armchair (COR, Germany)

1988 – Antares
Couch and day bed. Upholstery on aluminium frames (Ligne Roset, France)

1988 – Lumen*
Coffee table with indirect lighting. Maple and matte glass on a black metal frame (Ligne Roset, France)

1989 – Cristalli*
Bath tile series (Villeroy & Boch, Germany)

1990 – Taurus
Table and chair group. Beech and lacquer (COR, Germany)

1990 – Tarbo
Seating elements, with a design based on gentle curves (COR, Germany)

1990 – Edition Peter Maly
Limited edition design objects. Pear wood, high-gloss lacquer, maple, high-grade steel. »La Bandiere« and »Elliptic« (cupboards),«Thor« (table) and »Ivory« (secretary) (Intra Design, Switzerland)

1990 – Arcus
Bedroom program and bed linen collection. Printed satin (Domitex, Germany)

1990 – Rondo*
Cupboard with an arched front. Pear and lacquer (Reim Interline, Germany)

1990 – Hera
Chair made of formed plywood shells on a metal tube frame (Ligne Roset, France)

1990 – Agora
Round dining table with blue tinted glass on a domed shell. Supporting base made of wood and circular steel tubes. Black (Ligne Roset, France)

1990 – Duo
Functional container with hinged doors. Can be used as a bar, or to support a stereo or television. Wood and lacquer combination (Interlübke, Germany)

1990 – Zelda
Round tables with a lathe-turned central supporting column. Adhesive bonded cherry and maple (COR, Germany)

1990 – Axis*
Cupboard and wall panelling system (Poliform, Italy)

1990 – Contur*
Couch and three different individual armchairs (Wittman, Germany)

1991 – Edition Peter Maly
»Fluctus« – cupboard with centrally mounted folding doors. Maple and pear wood stained grey, with veneer strips. »Torii« – lacquered shelves with pear wood facing throughout (Intra Design, Switzerland)

1991 – Tarbo
Upholstered bed with a bevelled frame (COR, Germany)

1991 – Onde
Bedroom cupboard system. Postformed front, lacquered or with a beechwood veneer. Beech dividers (Ligne Roset, France)

1991 – Boa
Seating unit system with serpentine backrests (Gelderland, the Netherlands)

1991 – Koga*
Seating elements with high back and armrests (Ligne Roset, France)

1991 – Tectus
Dining room chairs of solid cherry (Bros's, Italy)

1991 – Schach
Dining room table with chequered veneer and inlays (Bros's, Italy)

1991 – Bibo
Reading lamp which swivels and tilts. Nickel and glass (Anta, Germany)

1992 – The New Biedermeier
Unique sofa (Gelderland, the Netherlands)

1992 – Felix
Seating furniture group (Brunati, Italy)

1992 – Sharky
Adjustable couch with numerous seating positions (Brunati, Italy)

1992 – Vega (»Veganova« as of 1994)
Armchair and sofa with a zestful design (COR, Germany)

1992 – Metrix-Wohnen
Wall-to-wall cabinet system in a variety of materials for a number of different possible combinations (behr, Germany)

1992 – Arc
Comfortable easy-chair. Wood shell on a metal base. Upholstery: leather or Alcantara (Ligne Roset, France)

1992 – Quadro, Undine, Intarsia
Extension to the »Facades« solitaire cupboards. Pear wood with various inlays (Reim Interline, Germany)

1992 – Ventrio*
Wedge-shaped sideboard. Cherrywood and varnish combination (Reim Interline, Germany)

1992 – Cirrus
Combination seating/sleeping furniture with hinged backrest. Seat depth can be adjusted by sliding the base out; pulled out completely it becomes a bed. Black or aluminium-coloured metal structure. Leather or fabric upholstery (COR, Germany)

1992 – Pando
Light armchair with aluminium-coloured metal structure. Leather upholstery with a matt finish (COR, Germany)

1993 – Cycla
Line of table accessories: vases, bowls, candlesticks, coffee and tea service, trays, hurricane lamp, cruet stand. Silver or high-grade steel with pear wood (Wilkens Bremer Silberwaren, Germany)

1993 – Cuneo (together with Christian Werner)
Series of individual furniture pieces: cupboards, shelves, dining table and chairs. Solid cherry frames (Consonni, Italy)

1993 – Biblia
System consisting of bookcases, tables and side tables (Ligne Roset, France)

1993 – Papilio
Cupboard to supplement the Peter Maly Edition. Pear wood. Blue mirrored interior (Intra Design, Switzerland)

1993 – Peter Maly Teppichedition
»Ganymed«, »Gondwana«, »Galaxi«, »Crossing«, »Cubes«, »Centercourt«, »Genesis«, »Match«, »Monega«, »Medio I + II«, »Punto«, »Minipoint«, »Nirvana«, »Najade«, »Nomos«. 100% pure new wool, hand-made, inlaid with copper, brass, leather, silk, and high-grade steel (JAB Anstoetz-Teppiche, Germany)

1993 – Quadrat Nova
Extension to the »Quadrat« series. Cupboard and sideboard combination. Cherry frame, body lacquered in black (Reim Interline, Germany)

1994 – Mosaik
Supplement to the »Facades« series. Cupboard with a checkered veneer and black and white wood inlays; cherrywood and varnish (Reim Interline, Germany)

1994 – Thetis
Supplement to the »Facades« series. Cherrywood with metal inlays (Reim Interline, Germany)

1994 – Ticino
Dining room line: sideboards, glass cabinets, dining tables, upholstered chairs. Beech, linoleum, fabric, frosted glass (WK Wohnen, Germany)

1994 – Metrix-Schlafen
Bedroom furniture system: cupboard system with an exquisite interior. Upholstered bed, bed with slats, panel bed, bedside furniture, and bed linen (Behr, Germany)

1994 – Basic
Shelf system with panels and sliding doors. Cherry and aluminium (Reim Interline, Germany)

1994 – Sera
Lounge-chair and couch with adjustable arm rests. Nickel-plated steel frame, with leather/fabric upholstery (COR, Germany)

1994 – Cocon
Oval day bed and oval cherry table (COR, Germany)

1994 – Ludwigsburg
Wall-mounted lamp providing indirect light. Aluminium-coloured metal structure, lacquer (Anta, Germany)

1994 – Thonet 737 + 1737
Table and chair group, beech frame. Upholstered or covered with thick cotton strips. Table tops of beech and linoleum (Thonet, Germany)

1995 – Parallele
Chest of drawers, cupboards and beds on casters. Beech, lacquer, corrugated aluminium (Ligne Roset, France)

1995 – Habitus
Armchair with bent laminated-wood armrests. Beech, leather, fabric (Ligne Roset, France)

1995 – Incognito
Sofa with removable fabric cushion covers and visible zippers, on casters or feet (Ligne Roset, France)

1995 – Presto
Tray table on casters (Ligne Roset, France)

1995 – Les Points
Sisal rugs, with high-grade steel eyelets and tufts of wool (Ligne Roset, France)

1995 – Margo
Upholstered bed, exclusive bed linen, oval swivelling bedside trays. Printed satin, maple (Domitex, Germany)

1995 – Cena
Table and chair group. Chairs with upholstered or aluminium backrests in various forms. Beech supporting frame. Tables with a beech or metal frame, table tops of beech, linoleum, glass (Tonon, Italy)

1996 – Menos
Living and dining room furniture system. Unit elements in different colour velvety finishes. Shelves, panels and tables of maple or beech. (behr)

1996 – Mato
Modular seating elements, forming benches of solid wood with webbing. Loose chair cushions, arm- and backrests to be slotted in in various positions. Coordinated coffee tables with stone tops, and one-off armchairs with webbing. (COR)

1996 – Tameru, Tameras, Someru
Extension of the P.M. carpet collection. 100% pure wool, hand-tufted, with stainless steel and copper inlays (JAB Anstoetz)

1996 – Imago, Artes, Cura
Three pianos in contemporary designs. Imago: pearwood frame, satin finished body. Artes: black high-gloss finish, stainless steel inlays. Cura: cherry with stainless steel inlays. (Sauter)

1996 – Duo (re-design)
Revision and extension of 1984 purist design classic: new metal knobs/grips, aluminium frame/glass doors, shelf units, dining table. (Interlübke)

1996 – Rota
Chair family in various versions: solid wood, moulded or upholstered backrests. Round extendable dining table, cherry or beech. (Tonon, Italy)

1997 – Chess
Bed with maple posts and textile wall hanging, maple bedside table with coordinate bedlinen collection. (Domitex)

1997 – Top
Shelf system, easy to slot together. Horizontal lines, aluminium-colour metal parts, beechwood shelf boards. (ligne roset, France)

1997 – Swing
Armchair and sofa in a sculptural design. Aluminium-colour metal frames. (ligne roset, France)

1997 – Ellipse
Three lamps with buttoned-on lampshades of polycarbonate fibre material. (ligne roset, France)

1997 – Tecton
Unit furniture: ›System‹ family redesign. Square-grid skeleton, with various slide-in front elements. Large-format sliding doors to shut functional areas. (behr)

1997 – Vivace
Grand piano in contemporary design. Black high-gloss finish, or maple veneer, stainless steel leg caps/ wheels and decorative inlays. (Sauter)

1997 – Chimney hooks
Various hooks in stainless steel with leather grips, wall bracket or floor stand. (Conmoto)

1997 – Alpha
Dining-room table and chairs, solid beech, maple or cherry. Table frame: solid wood or stainless steel. (Tonon, Italy)

1997/98 – Les Carrées
Quadri, Bayadere and Sendai: one-off carpets of 100% pure wool. (ligne roset, France)

1998 – Carrée
Bedlinen collection, pure cotton, different colours, with checked edges. (Domitex)

1998 – Circum
Couch with pull-out seat cushions to make a chaise-longue. Metal frame with nickel-colour coating, leather or fabric covering. (COR)

1998 – Circo
Revolving chair, light and circular, upholstered seat, backrest available upholstered or webbing. Metal frame with nickel-colour coating, leather or fabric covering. (COR)

1998 – Onda
Piano in minimalistic design. Wave-textured front, aluminium varnish and maple. (Sauter)

1998 – Parallele panel/shelf system
Low natural beech panels with horizontal slits to slot in metal aluminium-coated boards. (ligne roset, France)

1998 – Screen
Lamp family: Large and small table lamps, floor lamp, wall sconce. Frosted glass, mat-gloss chromium-plated metal frames. (Anta)

1998 – Modul 5000
Office desk line with multiple variations. Cast-aluminium frames with diverse platforms and different materials. (Mauser Office)

Mitarbeiter Staff List

Für einen erfolgreichen Designer und Innenarchitekten wäre eine Bewältigung der Aufträge und Projekte, die an ihn herangetragen werden, unmöglich, könnte er nicht auf einen eigenen Stab von festen und freien Mitarbeitern zurückgreifen.
Im folgenden sind die Namen derer genannt, die Peter Maly zur Zeit unterstützen oder in der Vergangenheit unterstützt haben.

A successful designer and interior designer could never cope with all the contracts and projects he is involved in without being able to rely on the assistance either of a team of staff members or the services of freelancers.
The following therefore provides a list of those who have supported or continue to support Peter Maly in his work.

Arrangements/Styling

Gisela Maly (seit 1974)

Industriedesign

Jens Ahlemeyer (1993-1995)
Antje Härtel (1993-1996)
Christian Werner (1987-1992)

Innenarchitektur

Renate Dipper (1988-1990)
Heike Endlich (seit 1991)
Harry Grieshaber (1971-1975)
Elmar Janik (1990-1991)
Siw Matzen (seit 1996)
Ursula Rettig (1987-1988)
Franz Schulte (1976-1978)
Bernd Benninghoff (seit 1996)
Stefan Thielecke (1995-1996)
Hartmut Tromberend (1978-1987)

Office Management

Karin Görling (seit 1985)

Settings/Styling

Gisela Maly (as of 1974)

Industrial Design

Jens Ahlemeyer (1993-1995)
Antje Härtel (1993-1996)
Christian Werner (1987-1992)

Interior Design

Renate Dipper (1988-1990)
Heike Endlich (as of 1991)
Harry Grieshaber (1971-1975)
Elmar Janik (1990-1991)
Siw Matzen (as of 1996)
Ursula Rettig (1987-1988)
Franz Schulte (1976-1978)
Bernd Benninghoff (as of 1996)
Stefan Thielecke (1995-1996)
Hartmut Tromberend (1978-1987)

Office Management

Karin Görling (as of 1985)

Firmenadressen Index of Firms

Eine umfassende Darstellung aller von Peter Maly gestalteten Objekte kann und will dieses Buch nicht leisten. Um weitere Informationen über die hier vorgestellten Produkte zu erhalten, kann oft auf Herstellerinformationen und Prospektmaterial zurückgegriffen werden. Da zudem viele der vorgestellten Objekte noch erhältlich sind, soll interessierten Lesern mit dieser Liste auch die Möglichkeit gegeben werden, sich über Bezugsquellen zu informieren.

It is not, nor could it be the intention of this book to provide a comprehensive presentation of all the products Peter Maly has designed. Further information on the objects described above can often be supplied by the manufactzrer in the form of brochures and leaflets. Given that many of the products presented here are still available, the following list is meant to assist readers in applying directly to the manufacturers for additional information.

Anta Leuchten GmbH
Osterbrooksweg 59
22869 Schenefeld
Deutschland/Germany

Behr International
Möbelfabrik Erwin Behr GmbH & Co.KG
Behrstraße 100
73237 Wendlingen
Deutschland/Germany

Bros's srl
via Catalani, 5
20035 Lissone
Italien/Italy

Brunati Italia srl
via Nazionale, 61
33048 S. Giovanni al Natisone
Italien/Italy

Conmoto
Wagner-Kollektion
Lümernweg 188 A
33378 Rheda-Wiedenbrück
Deutschland/Germany

COR Sitzmöbel
Lübke GmbH & Co
Nonenstraße 12
33378 Rheda-Wiedenbrück
Deutschland/Germany

Domitex International
Nickelstraße 1
33378 Rheda-Wiedenbrück
Deutschland/Germany

Gelderland meubelindustrie b.v.
Stationsweg 7
4100 AA Culemborg
Niederlande/The Netherlands

Interlübke
Gebr. Lübke GmbH & Co. KG
Ringstraße 145
33378 Rheda-Wiedenbrück
Deutschland/Germany

Intra Design AG
Letziweg 9
4663 Aarburg
Schweiz/Switzerland

JAB Anstoetz Teppiche
Dammheider Straße 67
32052 Herford
Deutschland/Germany

Reim Interline
seit/ as of 1998:
Interior Design Christine Kröncke
Thierschstraße 37
80538 München/Munich
Deutschland/Germany

Ligne Roset
b.p.no 9
01470 Briord
Frankreich/France

Mauser Office GmbH
34513 Waldeck

Oldenburger Möbelwerkstätten GmbH
Sanderstraße 21
49413 Dinklage
Deutschland/Germany

Richter Möbelwerkstätten GmbH
Von Tambach-Straße 2
49584 Fürstenau
Deutschland/Germany

Sauter
Pianofortemanufactur GmbH & Co. KG
Max-Planck-Straße 20
78549 Spaichingen
Deutschland/Germany

Tecta Möbel KG
Sohnreystraße 10
37697 Lauenförde
Deutschland/Germany

Thonet
Gebrüder Thonet GmbH
Michael-Thonet-Straße 1
35059 Frankenberg
Deutschland/Germany

Tonon S.p.A. & C.
via Diaz, 22
33044 Manzano
Italien/Italy

Wilkens Bremer Silberwaren AG
(seit/as of 1995: Bremer Silberwaren Handelsgesellschaft mbH)
An der Silberpräge
28305 Bremen
Deutschland/Germany

WK Wohnen
Einrichtungs-GmbH
Heilbronner Straße 4
70771 Leinfelden-Echterdingen
Deutschland/Germany

Fotonachweis Photo Credits

Es wurden von Seiten des Verlags alle Anstrengungen unternommen, um die Copyright-Inhaber des in diesem Buch verwendeten Bildmaterials ausfindig zu machen. Für Auslassungen entschuldigen wir uns und würden uns freuen, in künftige Auflagen die richtigen Angaben aufnehmen zu können.
Hinweise bitte an: Verlag form GmbH, Hanauer Landstraße 161, 60314 Frankfurt am Main.
Die Reihenfolge der Nennungen entspricht der Abfolge der Bildunterschriften auf den angegebenen Seiten.

All due effort has been made on the part of the publisher to find out who held the copyright to the various photos displayed in this book. We would like to apologize in advance should we have left someone out, and would be happy to include the correct entries in later editions. Please send information to: Verlag form GmbH, Hanauer Landstrasse 161, 60314 Frankfurt am Main, Germany.
The sequence of the names corresponds with the sequence of picture captions on the respective pages.

S./p. 17: Richard Stradtmann für »Schöner Wohnen«
S./p. 19: Richard Stradtmann für »Schöner Wohnen«
S./p. 20: Richard Stradtmann für »Schöner Wohnen«
S./p. 21: Richard Stradtmann für »Schöner Wohnen«
S./p. 22: Richard Stradtmann für »Schöner Wohnen«
S./p. 23: Richard Stradtmann für »Schöner Wohnen«
S./p. 24: Richard Stradtmann für »Schöner Wohnen«
S./p. 25: Richard Stradtmann für »Schöner Wohnen«
S./p. 26: Richard Stradtmann für »Schöner Wohnen«
S./p. 28: Richard Stradtmann
S./p. 29: Richard Stradtmann
S./p. 30: Richard Stradtmann
S./p. 31: Richard Stradtmann
S./p. 33: Rudolf Schmutz jr.
S./p. 34: Rudolf Schmutz jr.
S./p. 35: Achim Deimling-Ostrinsky
S./p. 35: Wolf P. Prange (Porträt)
S./p. 36: Rudolf Schmutz jr.
S./p. 37: Rudolf Schmutz jr.
S./p. 38: Rudolf Schmutz jr.
S./p. 39: Wulf Brackrock
S./p. 40: Rudolf Schmutz jr.
S./p. 41: Rudolf Schmutz jr.
S./p. 42: Hajo Willig, Frank Taeger, Stefan Thurmann, Ligne Roset
S./p. 43: Hajo Willig, Frank Taeger, Stefan Thurmann, Ligne Roset
S./p. 44: Rudolf Schmutz jr.
S./p. 45: Rudolf Schmutz jr., Amos Schliack
S./p. 46: Rudolf Schmutz jr.
S./p. 47: Bob Lienders
S./p. 48: Rudolf Schmutz jr.
S./p. 49: Rudolf Schmutz jr., Amos Schliack
S./p. 50: Rudolf Schmutz jr.
S./p. 51: Studio Armando Rebatto
S./p. 52: Stefan Thurmann
S./p. 53: Stefan Thurmann
S./p. 54: Hayo Heye
S./p. 55: Hayo Heye
S./p. 57: Stefan Thurmann
S./p. 58: Stefan Thurmann
S./p. 59: Stefan Thurmann
S./p. 60: Stefan Thurmann
S./p. 61: Stefan Thurmann
S./p. 62: Stefan Thurmann
S./p. 63: Stefan Thurmann
S./p. 64: Stefan Thurmann
S./p. 65: Stefan Thurmann
S./p. 75: Till Leeser
S./p. 76: Rudolf Schmutz jr.
S./p. 77: Rudolf Schmutz jr.
S./p. 78: Rudolf Schmutz jr.
S./p. 79: Rudolf Schmutz jr.
S./p. 80: Foto Frutig
S./p. 81: Rudolf Schmutz jr. für »Schöner Wohnen«
S./p. 82: Helge Mundt
S./p. 83: Amos Schliack
S./p. 84: Helge Mundt
S./p. 85: Helge Mundt
S./p. 86: Frank Taeger, Stefan Thurmann
S./p. 87: Frank Taeger, Stefan Thurmann
S./p. 88: Hayo Heye
S./p. 89: Rudolf Schmutz jr.
S./p. 90: Hayo Heye
S./p. 91: Miro Zagnoli
S./p. 92: Hayo Heye
S./p. 93: Till Leeser
S./p. 94: Frank Taeger
S./p. 95: Frank Taeger
S./p. 96: Hayo Heye
S./p. 97: Stefan Thurmann
S./p. 98: Stefan Thurmann
S./p. 99: Stefan Thurmann
S./p. 101: Hayo Heye
S./p. 102: Hayo Heye
S./p. 103: Hayo Heye
S./p. 104: Hayo Heye
S./p. 106: Rudolf Schmutz jr.
S./p. 107: Rudolf Schmutz jr.
S./p. 108: Hayo Heye
S./p. 109: Hayo Heye
S./p. 110: Stefan Thurmann
S./p. 112: Wolfgang Wilde
S./p. 114: Mercedes-Benz
S./p. 115: Mercedes-Benz
S./p. 117: COR
S./p. 132: Joachim Schumacher (P. Zec)
S./p. 133: Herlinde Koelbl (H. Lübke)
S./p. 119: Rudolf Schmutz Jr. (Cor-Sofa)
Stefan Thurmann (behr-Kommode)
S./p. 120: Amos Schliak
S./p. 121: Stefan Thurmann
S./p. 122: Rudolf Schmutz Jr.
S./p. 123: Hayo Haye
S./p. 124: Hayo Haye
S./p. 125: Fotoatelier Kaspar
S./p. 126/127: Rudolf Schmutz Jr.
S./p. 128/129: Bent Weber
S./p. 130 oben/above: Hayo Heye
S./p. 130 unten/below: Rudolf Schmutz Jr.
S./p. 131 oben/above: Hayo Heye
S./p. 131 unten/below: Bent Weber
S./p. 132/133: Peter Maly
S./p. 134: Bent Weber
S./p. 135: Rudolf Schmutz Jr.
S./p. 136: Rudolf Schmutz Jr.

Umschlag Vorderseite/front cover:
Bent Weber (oben/above)
Stefan Thurmann (Mitte/center)
Rudolf Schmutz Jr. (unten/below)
Umschlag Rückseite/back cover:
Christoph Schneider
linker Klappentext/front inside cover:
Rudolf Schmutz Jr.
rechter Klappentext/back inside cover:
Daniel Stier
Buchrücken/spine: Rudolf Schmutz Jr.

Register

Mit folgender Liste soll dem Leser ein Mittel an die Hand gegeben werden, Personen, Produkte und Projekte, die im Buch genannt oder gezeigt werden, schnell zu finden und die Monographie auch als Nachschlagewerk zu nutzen. Zudem werden über diesen Index die über das ganze Buch verteilten Statements erschlossen.
Kursiv gesetzte Zahlen weisen auf Abbildungen hin.

Index

The following list is designed to facilitate the use of this monograph as a reference work, enabling the reader to find quickly the persons, products and projects mentioned or shown in this book. This index also serves as a key to the statements to be found at intervals throughout the book. Numbers in italics refer to illustrations.

Über die Autoren

Verner Panton,

geboren 1926 in Gamtofte auf Fünen/Dänemark, studierte Architektur an der Technischen Hochschule in Odense, an der Kunstakademie Kopenhagen sowie an mehreren anderen europäischen Universitäten.
Nach Beendigung seines Architekturstudiums arbeitete er von 1950 bis 1952 für Arne Jacobsen. 1955 ließ er sich in der Schweiz nieder und arbeitet seitdem als selbständiger Architekt und Designer für Firmen wie »Unika Vaev«, »Kvadrat«, »Pluslinje«, »Mira-X«, »Louis Poulsen«, »Royal Copenhagen«, »Thonet« und »Vitra«.
Verner Panton schuf die »Visiona« 1968 und 1970 sowie die Kantine des Spiegel Verlags und die des Verlagshauses Gruner + Jahr.
Sein wohl bekanntestes Möbel, der Vollplastikstuhl »Side Chair«, ist eines der zentralen Symbole der sechziger Jahre.
Verner Panton starb 1998.

Josef Kremerskothen,

geboren 1925 in Iserlohn, studierte nach einer Tischlerlehre Architektur an der Technischen Hochschule in Karlsruhe. Gemeinsam mit Egon Eiermann entwarf er unter anderem den deutschen Pavillon für die Weltausstellung 1958 in Brüssel und gewann den Wettbewerb für die Gedächtniskirche in Berlin.
Ab 1956 war er in einer Architektengemeinschaft in Frankfurt am Main tätig. 1961 ließ er sich in Hamburg nieder und wechselte in die Redaktion von »Schöner Wohnen«. 1969 übernahm er die Chefredaktion dieser Zeitschrift, der er nahezu zwanzig Jahre – bis 1987 – vorstand.
Der Architekt und Journalist ist Kursusleiter an der »Henri-Nannen-Schule« in Hamburg.

Verner Panton

Peter Zec

Peter Zec,

der 1956 in Osnabrück geboren wurde, studierte dort Medienwissenschaft, Psychologie und Kunstwissenschaft und promovierte 1986 über das Thema »Geschichte, Technik und Kunst der Holografie«.
Von 1985 bis 1986 war Peter Zec Leiter des Museums für Holografie und neue visuelle Medien in Pulheim.
Vor einer zweijährigen Lehrtätigkeit als Dozent für Medien- und Kulturtheorie an der Universität Karlsruhe (1988/89), leitete er bis 1988 den Fachbereich »Bild« der Projektgruppe des Zentrums für Kunst- und Medientechnologie (ZKM) in Karlsruhe. Von 1989 bis 1991 Geschäftsführer des Verbandes Deutscher Industrie-Designer (VDID) und des Bundes Deutscher Grafik-Designer (BDG), leitet er seit 1991 das Design Zentrum Nordrhein Westfalen in Essen.
Im Oktober 1993 nahm Zec seine Lehrtätigkeit als Professor für Wirtschaftskommunikation an der Fachhochschule für Technik und Wirtschaft in Berlin auf.
Er ist Verfasser und Herausgeber mehrerer Bücher zu den Themen Neue Medien und Design.

Wolfgang Nagel,

1949 in Hamburg geboren, studierte in seiner Heimatstadt Literaturwissenschaft.
Von 1976 bis 1981 war er Redakteur beim »ZEIT-Magazin«, anschließend zwei Jahre Kulturreporter beim »Stern«. 1983 machte er sich mit einem eigenen Redaktionsbüro selbständig und beschäftigte sich in Beiträgen für das »F.A.Z.-Magazin«, »Geo«, »Manager Magazin« und andere vor allem mit Themen aus den Bereichen bildende Kunst, Design und Architektur.
1992 trat er in die Redaktion von »Architektur & Wohnen« ein; 1993 übernahm er die Chefredaktion dieser Zeitschrift.

Helmut Lübke,

der 1936 in Rheda geboren wurde, gründete nach seiner Tischlerlehre, einer kaufmännischen Ausbildung und einer Weiterbildung zum Betriebstechniker 1954 gemeinsam mit einem Partner die Polstermöbelfabrik Helmut Lübke und Co. (heute COR). 1960 übernahm er die Geschäftsleitung des Unternehmens, die er Ende 1994 an seinen Sohn übergab. Seit Juni 1996 ist Helmut Lübke Geschäftsführer beim Möbelhersteller interlübke.
Helmut Lübke ist Vorsitzender des Fachbeirats der Internationalen Möbelmesse Köln sowie Präsident des Verbands der Deutschen Möbelindustrie.

Wolfgang Nagel

Petra Schwab,

1956 in Gießen geboren, schloß 1983 in Hamburg als Diplom-Textildesignerin ihr Studium an der Hochschule für bildende Künste ab.
Von 1985 bis 1992 arbeitete sie in der Redaktion der Zeitschrift »Schöner Wohnen«, zuletzt als Redakteurin für Design. Nach eineinhalbjähriger Tätigkeit als Redakteurin bei der Zeitschrift »Architektur und Wohnen« begann Petra Schwab an einem Konzept mitzuarbeiten, das die Fachzeitschrift des Rat für Formgebung, »Design Report«, tauglich für den Verkauf am Kiosk machte. Drei Jahre war sie Chefredakteurin des »Design Report« und ging danach als stellvertretende Chefredakteurin zur Publikumszeitschrift »Ambiente«.
Seit 1998 entwickelt Petra Schwab in ihrem eigenen Büro Themen für diverse Zeitschriften und schreibt als Kennerin der Branche vorrangig über Designbelange.

About the Authors

Verner Panton

was born in 1926 in Gamtofte on the island of Fyn in Denmark. He studied architecture at the Odense Polytechnic, at the Art Academy in Copenhagen, and at various other European universities.
After graduating as an architect, he worked for Arne Jacobsen from 1950 to 1952. In 1955, he moved to Switzerland and since then has worked as a freelance architect and designer for companies such as »Unika Vaev«, »Kvadrat«, »Pluslinje«, »Mira-X«, »Louis Poulsen«, »Royal Copenhagen«, »Thonet« and »Vitra«.
Verner Panton created the »Visiona« of 1968 and 1970, and designed the canteens at the publishing companies Spiegel and Gruner + Jahr.
Probably his most famous furniture design is »Side Chair«; made completely of plastic it is one of the central symbols of the Sixties.
Verner Panton died in 1998.

Josef Kremerskothen

Josef Kremerskothen

was born in 1925 in Iserlohn. After completing his training as a cabinetmaker, he studied architecture at the Karlsruhe Polytechnic. Together with Egon Eiermann, he designed, among other things, the German pavilion for the World Fair in 1958 in Brussels, and was commissioned to design the memorial church in Berlin.
In 1956, he started work in an architectural group in Frankfurt am Main. He moved to Hamburg in 1961, where he joined the editorial staff of the interior decor magazine »Schöner Wohnen«. He became editor-in-chief of the same publication in 1969, which he headed for nearly 20 years – until 1987.
He currently teaches in his capacity as architect and journalist at the Henri Nannen school in Hamburg.

Peter Zec

was born in 1956 in Osnabrück, where he later graduated in media studies, psychology and art. In 1986, he was awarded a doctorate for his thesis on »The history, technology, and art of holography«.
From 1985 to 1986, Peter Zec was the director of the Museum for Holography and New Visual Media in Pulheim. Prior to a two-year appointment as lecturer for media and culture theory at the University of Karlsruhe (1988/89), he was head of the »Image« section of the project group run by the Center for Art and Media Technology in Karlsruhe. From 1989 to 1991, he acted as managing director of Verband Deutscher Industrie-Designer (association of German industrial designers) and Bund Deutscher Grafik-Designer (association of German graphic designers). Since 1991, he has been at the helm of the Design Center of North Rhine-Westphalia in Essen.
In October 1993, Zec was appointed to the chair of economic communications at the Vocational College for Technology and Economics in Berlin. He is the author and editor of a number of books on new media and design.

Wolfgang Nagel

was born in 1949 in Hamburg, where he later studied literature.
From 1976 to 1981, he was a member of the editorial staff of »ZEIT-Magazin«. The subsequent two years were spent reporting on cultural matters for the magazine »Stern«. In 1983, he opened his own editorial office and started producing pieces for publications such as »F.A.Z.-Magazin«, »Geo«, »Manager-Magazin«, among others, focusing in particular on the visual arts, design and architecture.
In 1992, he joined the editorial department of »Architektur & Wohnen«, and became editor-in-chief of the same magazine in 1993.

Helmut Lübke

was born in Rheda in 1936. After apprenticeships first as a cabinetmaker and later in commerce, he trained as a technical operations manager. In 1954, together with a partner, he founded Helmut Lübke und Co., manufacturers of upholstered furniture and today known as COR, becoming managing director of the company in 1960. In 1994 he handed over these duties to his son. Since June 1996 he has been managing director of interlübke, manufacturers of furniture. Helmut Lübke is Chairman of the Special Advisory Council of the International Furniture Trade Fair in Cologne, and President of Verband der Deutschen Möbelindustrie (association of the German furniture industry).

Petra Schwab

was born in 1956 in Giessen, studied textile design at Hamburg Hochschule für Bildende Künste (fine arts academy) and graduated 1983 with textile design diploma.
1985-92 worked as an editor with the home-living magazine, »Schöner Wohnen«, in the end as editor in charge of design. Schwab then worked for 18 months as an editor of the magazine »Architektur und Wohnen« and after that started working with others on making »Design Report«, the official organ of the Rat für Formgebung [Design Council], marketable via newsdealers. She served as the magazine's editor-in-chief for three years before joining »Ambiente« als deputy editor-in-chief.
In 1998 Petra Schwab established herself as an independent journalist, writing on a variety of subjects – mainly on her specality, design – for a number of magazines and publications.

Petra Schwab

Helmut Lübke

Impressum Imprint

ISBN 3-931317-91-9

Printed in Germany.

Gestaltung/Design:
Heine/Lenz/Zizka, Frankfurt am Main
Lektorat/Editorial Department:
Anja Münker-Nanke
Übersetzung ins Englische/Translation
into English:
Rebecca Wallach, Annette Wiethüchter
Litho/Lithography:
Eichhorn.Frankfurt, Frankfurt am Main

Die Deutsche Bibliothek –
CIP-Einheitsaufnahme
Peter Maly/hrsg. von Alex Buck und
Matthias Vogt. Josef Kremerskothen…
[Übers. ins Engl.: Rebecca Wallach,
Annette Wiethüchter]. – Erw. Neuaufl. –
Frankfurt am Main: Verl. Form, 1998
(Designermonographien; 5)
ISBN 3-931317-91-9
NE: Maly, Peter [Ill.]; Buck, Alex [Hrsg.];
Kremerskothen, Josef; GT